THE CHANGING CHICKEN

Jane Dixon is Fellow at the National Centre for Epidemiology and Population Health at the Australian National University. She was co-editor of *The Social Origins of Health and Well-being* (2001), and is a member of the Australasian Agri-Food Research Network and the Food and Nutrition Special Interest Group of the Public Health Association of Australia.

THE CHANGING CHICKEN

CHOOKS, COOKS AND CULINARY CULTURE

Jane Dixon

A UNSW Press book
Published by
University of New South Wales Press Ltd
University of New South Wales
UNSW Sydney NSW 2052
AUSTRALIA
www.unswpress.com.au

First published 2002

National Library of Australia
Cataloguing-in-Publication entry:

Dixon, Jane, 1951– .
The changing chicken: chooks, cooks and culinary culture.

Bibliography.
Includes index.
ISBN 0 86840 477 2.

1. Poultry as food — Australia. 2. Chicken industry — Australia.
3. Food — Social aspects — Australia. I. Title.

641.365

Printer Griffin Press

CONTENTS

PREFACE

As the Bicentenary of white settlement descended upon Australia in 1988, questions of national identity assumed some urgency. This was not surprising. What was surprising was the emphasis in the popular media of linking identity to food, given the country's relatively invisible culinary tradition. Common refrains concerned the extent to which Australia's cuisine and identity reflected its British colonial heritage, our burgeoning multicultural hues, and the possible overshadowing of this emergence by the Americanisation of diets. People sought out the foods of indigenous Australians as a point of reference and considered how our diets might have evolved had we been settled by the French or the Dutch. One broad consensus emerged: Australians, unlike our contemporaries, seek out and combine fresh ingredients in a way not bound by any particular culinary code.

In a short space of time, Australians have become obsessed with food. And it seems that we share our multifaceted concerns with many other Western countries. How safe is our food supply? Should we allow GM foods to be grown? Does the explosion in obesity amount to a public health disaster or does it mark a phase of economic development? What does the Uruguay Round of GATT mean for the national economy and farmers? What should we eat tonight? Can we let the kids have a fast food meal every now and then? These last two questions are probably the most pressing wherever food is abundant. As Bannerman (1998) puts it, '[Australia's] real food culture has little to do with ideals and public debate. It is the cookery of ordinary households, as practised in suburban kitchens on ordinary weeknights, after a tiring day at work'.

My own food obsessions were played out similarly within the inward and outward looking emotions and behaviours that are inevitably unleashed by food. I became an early member of the Italian-based Slow Food Movement, had a cholesterol check, turned my inner city courtyard into an edible garden, ate in cafes at every opportunity to learn what was new, watched every episode of the BBC's 'Two Fat Ladies' to reclaim what I assumed to be my culinary heritage, and decided that culinary cultures would be a perfect way of remaining engrossed in doctoral studies. This book results from that period.

The study began when I decided to investigate the powerful relationships and processes that underpin the production, distribution and consumption of a popular food. The resulting material operates at three levels: it provides factual information about the everyday commodity of chicken meat; it adds to our theoretical understanding of how food systems are organised in contemporary western societies; and it provides a new methodological framework for the study of food and other commodities. In so doing, I am convinced of two things: that ordinary objects provide extraordinary insights into the concept of power, and that an interdisciplinary approach is the only way of capturing the poly-dimensional nature of power.

Many people have contributed to the research. The first to do so were the nineteen respondents to a questionnaire soliciting the commodity to be studied: little did they know that they would set up an obsession with chooks. The second group includes those who spent time being interviewed and then reading and returning the transcripts. I am particularly appreciative of the following for spending many hours tutoring me on the workings of poultry production: Louis Vorstermans, Spencer Field, Rod Fenwick and Brian Johanning. Jeff Fairbrother, Gis Marven, Chris Turner, Johanna Barker, Tim Luckhurst, Geoff McGeachie and Wally Shaw provided frank advice and conflicting views about the balance of power. Adrian Simon and Jim Sclater revealed the basics of food product development. Ian Milburn and Chris Ellis offered insights about the struggles endured by the small, alternative producer, while John Cester provided material on the lot of the small retailer. Tony de Thomasis and Gisela Weeser of Safeway/Woolworths and Peter Presnell, Teresa Tremelling and Peter Jackson of Coles Myer were also generous with time and material. I am grateful to both supermarket chains for providing me with data that was not on the public record. The National Union of Workers and the Centre for Workplace Culture Change shared their material and, along with Gerald Lafferty, helped me to understand the labour process on the shop floor. And I am indebted to the thirty-three members of the five focus groups for their enthusiastic participation in the fieldwork.

Over the seven year project, additional support and help were forthcoming from numerous quarters. First, there were patient friends including Annie Ackland-Prpic, Lesley Hoatson, Gill McBride, Wendy Weeks and Lu and David Wilson. Tessa Morrison presented me with a chicken etching and Ruth Crow fed me with produce from her balcony. The forbearance of my work colleagues was vital and I was fortunate with my associates in the School of Social Science and Planning at the Royal Melbourne Institute of Technology, and the National Centre for Epidemiology and Population Health at the Australian National University. A reference point for the overall research approach was supplied by colleagues in the Australasian AgriFood Research Network and by Colin Sindall, who was engaged in the world of food and nutrition policy. Consultations with other experts, including John Burgess, Pat Crotty, Kim Humphery, Anitra Nelson, Michael Symons, Max Watson and Bev Wood, Wally Seccombe, Alan Warde and Bill Friedland were truly appreciated. My doctoral supervisors, Pavla Miller and Belinda Probert, offered unconditional support, openness to strange lines of enquiry, and diligence in keeping me focussed. Pippa Carron came to the rescue with proofreading of an earlier version, while the thoughtful and meticulous copyeditor Cassie Futcher and Susan Lindsay provided editorial and typing assistance on the final product. I am also grateful for the contribution of the Australian National University towards the cost of this publication. Finally, family members — Pat, David and Jo, George and Ivy — have endured hostile sights, sounds and absences, while the memory of Pete and Aberdeen Sausage was a guiding force. My biggest thanks goes to Colin, who was continually prepared to confront the practical and conceptual complexities, generally over wine and a meal.

Food may pose very mundane concerns — availability, acceptability and safety — but the table chicken reminds us of the profound underpinnings of modern food systems. Further, the demise of the chook shed represents the passage from a familiar food supply to a food system riddled with uncertainty and alienation. Food as fuel has been displaced by fear of food or, put another way, concerns about availability have been sidelined by issues of acceptability. Clearly, what is acceptable in a culinary culture is highly debatable and I hope that this book contributes to that debate.

Jane Dixon
Canberra

1

BARBECUES, CHICKEN SHEDS AND CULINARY DYNAMISM

Throughout the 1990s, many Australians witnessed a debate in the popular media about the existence or otherwise of an Australian culinary culture. Do we have one? What are its features? How is it distinctive? As part of the search for what was special about the Australian diet more universal questions were being asked. Is a culinary culture about unique food items, such as indigenous foods? Does it concern peculiar practices, such as the rapid adoption of new foods or the reformulation of traditional dishes? Can it be characterised by the dominance of a particular type of food supply, namely mass production or craft production? Is it about the way food is situated in the national psyche?

Take the barbecue, for example. This cooking device which is located in most suburban backyards, in numerous town parks and on inner city balconies has become a potent symbol of Australian identity. We promote ourselves overseas with images of prawns/shrimps (depending on the audience) being 'thrown on the barbie'. Where many commentators have used the widespread enthusiasm for this particular form of cooking to judge Australia's culture as distinctively casual, another interpretation of Australians' relaxed relationship to food has emerged. One respected market researcher, who over the years has investigated how Australians use food to cope with the impact of social change, reported that we had been overwhelmed by a desire to relax (Mackay 1992). The researcher found that by the early 1990s *how* we eat counted as much to us psychologically as *what* we eat (Mackay 1992, p. 5). Creating opportunities for casual eating experiences appeared from his research to be almost obsessional. He states that against the backdrop of the redefinition of gender roles, marriage, families, politics, work, money, shopping and cultural identity and what he describes as generalised anxiety:

> ... it is perhaps inevitable that Australians would reach the point where they want to be let off the hook; where they want to experience some relief from pressure ... At the present moment, relaxation has emerged as the key word: 'coping' is seeming to growing numbers of Australians to be a matter of 'relaxing' (Mackay 1992, p. 6).

In this way, 'attitudes towards food and eating emerge as one of the most significant symbols of the process of adaptation [to economic and social changes]' and Mackay singles out the barbecue, or the informal outdoor alternative to the indoor kitchen, as a practice that reflects adaptation (Mackay 1992, p. 12). He adds that women welcomed the tradition of men cooking with the barbecue because barbecued (and take away) meals help women to 'buy time' in order to relax. On Mackay's evidence, relaxed eating appears to be a survival tactic rather than a casual spirit at work and from reading trade magazines, such as *Restaurant News*, does not seem unique to Australia.

Indeed, instead of unearthing any unique culinary culture, numerous commentators have raised concerns about the consumption of foods and practices endemic to the United States that seem poised to sprawl the world over. Gastronomes and even professors of nutrition comment upon the coca-colonisation of diets and the spread of the golden arches. In *Goodbye Culinary Cringe*, a celebration of Australia's new found freedom from the apron strings of its colonial gastronomic heritage, we were warned that an emergent Australian cuisine, based on fresh produce infused with multicultural cooking practices, could be readily undermined by an American influence. The book's author, Cherry Ripe, noted how houses in the United States were being built without kitchens and that home meal replacement was taking one in three food dollars (Ripe 1993, p. 130). She hinted that without vigilance we would follow, and sure enough, by the mid-1990s we were paying the same proportion to the food service sector to prepare previously home cooked meals (BIS Shrapnel 1995).

Personally touched by these debates, I was particularly interested by the reaction of my feminist friends to Cherry Ripe's portents. While they love food, especially handling fresh produce, pleasing others with their meals and socialising around food, nightly meal preparation is difficult: they have better things to do and besides, 'it's cheaper to buy meals than it is to prepare them'. To support this proposition they invariably cite the barbecue or rotisserie chicken. These cooked birds are cheaper than a bird in the raw state, and 'if you are really lucky with your timing at the supermarket' two small charcoaled 'chooks' can be purchased for the price of one.

The popularity of someone else's cooked chicken invited scrutiny, given the cultural significance of that bird's presence in the Australian social and physical landscape during the second half of the 20th century. Up unto the mid-1970s, to cook a chicken was a rare act even though many homes, both rich and poor, contained a basic shelter located in the back garden, called the chook shed, for rearing chickens and other poultry. In one food history, the suburban scene between the two world wars was described thus: '[m]any people's backyards contained rows of carrots, a lemon tree and maybe a loquat or quince or a passionfruit vine and mint growing by the rainwater tank. Scraps went to the "chooks"' (Symons 1982, p. 142). For most Australians over forty years of age the family chook yard, or the one over the back fence, was remembered for eggs and not for meat. Many city dwellers received a clutch of eggs from neighbours in exchange for a bag of apricots when there was a summer abundance and one of the delights, or chores, of being a child in the late forties and throughout the fifties was of feeding chooks before or after school. Without thinking ecologically the 'waste not, want not' mentality was a good provider.

Post World War 2 Australians were eating eggs daily for breakfast and were exporting them along with frozen chickens to food depleted Europe. Meat was consumed three times a day, but very little of this was from the backyard or farmyard chook, or the chooks kept at the nearby dairies for that matter. Only at Christmas and possibly Easter did the majority of Australians tuck into roast chicken. Considered a 'superior cut of meat', chicken was obtained from small specialist poultry shops in, or close to, the city markets, or from small family farms dotted around the cities. Roasting a chicken was an act of love, a symbol of a special occasion, a rare treat signalling a religious holiday.

For 175 years, from the time of the first white settlers, the few available food histories reveal that chicken consumption was negligible. By 1950 Australians had one of the highest intakes of meat in the world and it was a time of 'plentiful supplies of beef, mutton, lamb, rabbit, wheat and other grain crops unless drought or other disasters affected production or distribution' (Cahn 1977, p. 55). Chicken was not mentioned. However, Australia was also home to some extremely gifted bird breeders, including Norm Thomas. In 1946 he 'was granted the first export licence in South Australia for poultry to England, Hong Kong, Egypt and Arabia. Live birds were also sent to Borneo and Japan to feed prisoners of war' (Cain 1990, p. 109). By the 1960s, Thomas owned the Windsor Poultry Shop in Adelaide and he reportedly had a goal to provide all Australians with a regular supply of cheap chicken: '[t]he present price of 6/5 to 7/5 a lb puts chicken beyond the reach of most people as a regular Sunday dinner. It is my aim to bring down the price to a level at which people can afford chicken twice a week and make it comparative with the best beef' (Cain 1990, p. 111).

And he and the other avian breeders and hatcherymen succeeded. Yearly consumption in Australia soared from five kilograms per household in 1960 to twenty-eight kilograms in 1994. It now constitutes one quarter of our annual meat intake and, along with margarine, was the food to show the fastest growth in consumption since 1975 (Skurray & Newell 1993). We eat almost twice as much chicken as we do lamb and mutton and our legendary beef eating status is under threat. The once special meal for Anglo-Australians has now become a food of daily consumption for many. Chicken sandwiches are a lunchtime custom and dinners around the country often consist of a home cooked serving of chicken fillet covered in a sauce from a jar, most probably called Chicken Tonight. Roast chicken continues to be a favourite family meal, and two whole birds cooked on the supermarket rotisserie are snapped up at supermarket closing time. If rushed, tired or in need of a treat, an Asian inspired stir fry or a KFC bucket of crumbed drumsticks await not far down the road.

Daily consumption of chicken meat has been achieved despite the fact that the wood used to build the chook shed was thrown on the barbecue some years ago. Any sizzling chicken now comes from the poultry farms dotting the urban fringe, with only half a per cent of the chicken consumed in this country being home grown (ABS 1994).

Is it possible to learn something about culinary dynamism from the disappearance of one Australian icon, the emergence of another and a reliance on strangers to prepare and cook our foods? Does the evanescent chicken shed stand for a loss of home food production and the barbecue and its associations with consumer relaxation stand for consumers taking charge of their lives in an age of anxiety? Could the chicken offer an entry point to contemporary changes in culinary cultures and food systems?

In order to designate an appropriate food to sit centre of the plate and be gnawed at by sociological concepts, I wrote to twenty-six Australians and asked them to select the three most popular foods of the previous decade. Those whose opinions were sought included people considered to be setting the food industry and government agenda, as well as high profile chefs, nutritionists and a couple of 'noisy, opinionated' gastronomes said to be influencing producers' and consumers' behaviours. A remarkable consensus ensued from the nineteen respondents about the most popular food products. They nominated ready-to-eat breakfast cereals, chicken, and fast food popular with children, such as pizza. The responses from these knowledgeable *foodies* supported the official data. On this basis I decided that chicken would provide the foundation of a study on how culinary cultures and food systems change.[1]

EXPLORING CULINARY CULTURAL CHANGE

Once the matter of a food commodity was settled, I required a reference point to explore how western culinary cultures change. This very topic, it seems, is enjoying enthusiastic scrutiny in academia. Retail geography and three new branches of sociology have augmented social histories of food in the last two decades. The sociology of food has emerged to house the work of those studying the mutual organisation of food and social relationships. The sociology of consumption, which extends to an infinite list of goods and practices, is an area devoting significant energy to food. Thirdly, scholars in rural sociology and the 'new political economy of agriculture' are advancing our understanding of the so-called agrarian question.[2] The field of retail geography focuses on the demise of high street retailing, the planning issues associated with shopping malls, industry concentration among corporate retailers and the effect of the latter in transforming the countryside through their relationships with primary producers.

A quick scan of these fields reveals agreement that highly industrialised economies are experiencing some fundamental changes in relation to the food supply, in who is preparing the food, in what is considered a meal, and the place of food in sociality and culture. Some of the terms being used include the second food revolution, gastroanomie, the demise of the ritual meal and eating community, the disintegration of food regimes and new masters of the food system. These particular terms are introduced here to indicate the extensiveness of the changes in culinary cultures that are being canvassed.

Sokolov (1991), an examiner of gastronomy in the United States since European settlement, refers to contemporary changes of a revolutionary magnitude. Sokolov treads familiar ground when he depicts the first food revolution as triggered by the 16th century Spanish contact with the Americas and the cross-continental transportation of foods. He argues that a second food revolution is afoot, concerning a worldwide attitude to cuisine: namely the mixing of novelty and tradition. With widespread migration, tourism and cosmopolitanism each transmitting ideas about food, 'nouvelle cuisine' and regional 'ideocuisines' have emerged. New food service industries have grown around the prosperity of well-read and travelled urbanites. Furthermore, Sokolov suggests that the new cuisines are defeating the total dominance of supermarket cuisines. In a very different work, *Landscapes of Power*, Zukin supports Sokolov's contention that a new approach to food — including the entry of new food suppliers and different consumption norms — is growing around those gentrifying the cities of North America. 'Both gentrification and new cuisine represent a new organization of consumption that developed during the 1970s. Both imply a new landscape of economic power based, in turn, on changing patterns of capital investment, production and consumption' (Zukin 1991, p. 214). Part of the capital investment involves the employment of highly paid and celebrity chefs who join agricultural producers and a 'broad elite' to change cuisines. As a result, 'the consumption of nouvelle cuisine [is] spread not by military rule or cultural imitation but by market power' (Zukin 1991, p. 209).

Like Sokolov, the influential anthropologist Sidney Mintz (1994), uses the revolution metaphor to outline the nature of changes to contemporary food systems. Mintz describes the 1000-year old 'first revolution' as peasant and farmer-led through the domestication of plants and animals. In his opinion, the resulting food availability of the last 500 years in Western Europe has been overshadowed in the last one and a half centuries by an emphasis on two particular food products, namely processed sugars and fats. According to Mintz, national diets have been altered through the aggressive promotion of these industrial products, the impact of which has been the gradual erosion of the centuries-old intrameal structure that was made up by core-fringe-legume items.[3] The once marginal fats and sugars have displaced the complex carbohydrates which constituted the core of the meal and in so doing have 'altered the nature of the core itself — its dietary, nutritive and ideological contribution' (Mintz 1994, p. 112).

While Sokolov identifies the second revolution with the playful creation of diets, Mintz suggests that the social eater is being transformed as the significance of different foods is altered. While both grant power to different actors — in Sokolov's case to the consumer and for Mintz the corporate producer — their work converges on the point that contemporary culinary dynamism is less about available food products, despite the much celebrated, ever-expanding product range. The more important shifts involve the changing nexus between food products and food practices. The emphasis on the importance of food practices is supported by numerous social scientists who are alerting us to the breakdown in the intermeal structure, or the patterning of meal events. Practices of grazing, snacking, skipping meals and 'dashboard dining' are said to be upsetting the culinary order (Ripe 1993). Which foods are placed on the plate or in the bowl, how they are placed within daily and weekly lifestyle routines, and how they contribute to, and draw on, personal and social identities, are key issues for investigation.

The critical food practice of household cooking has been implicated in the current transformations occurring in culinary cultures. For Fischler, the French social scientist, '[c]ookery helps to give food and its eaters a place in the world, a meaning ... the culinary act sanctions the passage of food from Nature to Culture' (Fischler 1988, p. 286). Over the last two decades households have been relying more on industrialised food production and Fischler alleges that a reliance on industrial kitchens ushers in ignorance of foods, especially concerning food's attachment to nature. Consequently, food-based meaning systems become more tenuous and food-based rules become fluid. 'In a food system (and a cultural system) that is in the process of being destructured and/or restructured, how do we situate ourselves in the universe and cosmos?', asks Fischler (1988, p. 290). The dissolution of long accepted rules around eating results in a condition that Fischler has termed 'gastroanomie'.

Building on Fischler's insights, Falk reports on the passing of the eating community and of the ritual meal. According to Falk (1994) the introduction of privatised forms of eating, such as the take away snack, corresponds with the modern appearance of individuality. As eating is reorganised from communal to individual arrangements, so tastes are transformed. The social meal has been replaced in Falk's opinion by 'oral side-involvements' or 'oral pleasurables' involving snacks and substances 'not considered to be foods (sweets, titbits, soft and alcoholic drinks) or which actually fall outside the category of nutrition (tobacco, chewing gum)' (Falk 1994, p. 29).

At the same time that Falk and numerous others credit such changes with empowering consumers, questions are being asked about who else is benefited by the shift away from self-provisioning to reliance on the marketplace for foods. The purchase of goods and services in the market place, or commodification, has received significant attention from the self-proclaimed 'new political economy of agriculture'. Within this field many would agree with Friedmann's assessment that 'the purchased diet has become the means by which agro-food relations now encompass the globe, and ... penetrate ever more deeply into daily life' (Friedmann 1990, p. 193). In this context, agrofood relations refer to a growing concentration of a few global firms in key sectors exerting pressure on nation states and global regulatory institutions, like the World Trade Organisation (WTO).[4] As a result of their activities, corporate producers and traders influence norms of consumption more than consumers.

However, the emphasis on producer power within the sociology of agriculture is far from settled. Questions are being asked about the stability of agrofood relations, and specifically food regimes based upon 'links [between] international relations of food production and consumption [and] ... forms of accumulation' (Friedmann & McMichael 1989, p. 95). According to the food regimes theory, which is explained further in the next chapter, the world's food supply has been dominated by two food regimes over the last century, with the second one currently in crisis. There is a suggestion that a third food regime is coming into focus as a result of local groups finding alternatives to globalised food supplies (Friedmann 1993; LeHeron & Roche

1996). While a new pattern of events is awaited, consumer behaviour is entering into the deliberations of some of the most ardent political economists.

A quite different challenge to the idea that the balance of power in food systems emanates from the sphere of production comes from retail geographers and anthropologists, both of which study the intricacies of trading goods and services. According to one of the latter exponents, '[f]rom the social point of view, and over the span of human history, the critical agents for the articulation of the supply and demand of commodities have not been rulers but of course, traders' (Appadurai 1986, p. 33). What social histories of various commodities show is that merchants have not simply delivered goods to the marketplace but that they have delivered stories about the goods: they have imbued them with both mystery and relevance (Levenstein 1993; Mennell 1985; Mintz 1985). Merchants prefigured the value-adding activities undertaken by today's retailers and producers and in so doing built a bridge between producers and consumers.

According to retail geographers, the bridge for some commodities has become elongated through the process of commodification and the global trade in everyday, as opposed to luxury, foodstuffs. The activities of numerous buyers and sellers, along with a cacophony of falsehoods and truths about each and every food, lends a certain urgency to making foods desirable, or as Levi-Strauss (1978) put it, 'good to think in order to be good to eat'. What is becoming increasingly transparent is retail trading's two-way influence over production and consumption. Bourdieu, who has chronicled the reproduction of social status through food, remarks that retailing is:

> ... neither the simple effect of production imposing itself on consumption nor the effects of a conscious endeavour to serve the consumer's need, but the result of the objective orchestration of two independent logics, that of the fields of production and consumption ... the tastes actually realized depend on the systems of goods offered; every change in the system of goods induces a change in tastes. But conversely, every change in tastes ... will tend to induce ... a transformation of the field of production (cited in Ducatel & Blomley 1990, p. 216).

Retail traders, as the material in Chapters 6 and 7 reveal, are 'increasingly mediating the producer-consumer relation' (Lowe & Wrigley 1996). However, these entities, driven by the process of accumulating capital, are not alone in mediating this relationship. The celebrity chefs referred to earlier compete with nutrition educators and other food knowledge producers, advertisers, scientists and government bureaucrats in creating a pecking order of foods, such that consumers change their diets regularly. Located in what is known in economics as the spheres of distribution and exchange, these actors endeavour to counteract the gastroanomie that accompanies an industrialised food supply. They straddle the spheres of production and consumption by communicating meanings about foods. Furthermore, the way in which these charismatic state and professional authorities manufacture esteem for particular foods and re-embed trust in food supplies that are distanced geographically and metaphorically from consumers is of great value, symbolically and financially, to food producers and retailers alike. The large corporate firms among them are most enthusiastic about forging relationships with these actors in the middle.

In Australia, as elsewhere, there is increasing evidence that for a range of commodities and for the culinary culture as a whole, retailers are indeed 'the new masters of the food system' (Flynn & Marsden 1992). The two biggest supermarket chains are not simply producers and traders in goods and services but major cultural institutions of eighty years standing (Humphery 1998). It is they, not producers, that have given us ideas about convenience (in products as well as shopping), choice, thrift, cleanliness, order and more recently, family times, fun times and tradition by recreating market-type stalls filled with fresh produce. In short, corporate retailers are well situated to co-ordinate the relationships between producers and consumers by synchronising the cultural, economic and social values attached to the commodities they sell.

So while the different disciplines canvassed here agree that established food systems and culinary cultures are in a state of transition, they disagree over several fundamental issues. Major points of contention include the nature of the relationship between producers and consumers, how the balance of power is being exercised in an ever-more globalising world and how particular foods and food practices come to be esteemed in nations where food abundance co-exists with food anxieties and ignorance. This book aims to address these questions by systematically introducing a variety of perspectives of power into the study of food. The chief proposition is that culinary cultural change is not consensual, but is fought out in the marketplace of buyers and sellers through the practice of cultural and economic strategies. As Lipietz (1987) has pointed out in relation to the capitalist system, an absence of balance between production and consumption results in crises. In the case of a commodity, it seems reasonable to suppose that if production and consumption do not align then a faltering commodity complex follows. I propose that aligning cultural and material production is fundamental to shaping food tastes and to the dynamism in culinary cultures and food systems. Now the task is to take a table chicken in order to test the proposition.

THE SOCIAL LIFE OF COMMODITIES

Given that culinary dynamism and changes to food systems involve a great many actors, some rather complex interrelationships and a few hidden processes, care has to be taken to structure a study so that it provides both breadth and depth. As Chapter 3 reports, power in relation to food commodities has been examined for close to twenty years using the highly regarded Commodity Systems Analysis (CSA), developed by Friedland in 1984. My major concern with the CSA model as articulated then was its orientation to the contours of economic power. The cultural power of economic actors and the power relations embedded in the everyday lives of consumers could not be captured. For this reason, I was attracted to the works of Appadurai (1986) and Kopytoff (1986), who provide principles for studying power through a commodity in a way which helps to overcome a mono-disciplinary focus.

In Appadurai's estimation '[f]ocusing on the things that are exchanged, rather than simply on the forms or functions of exchange, makes it possible to argue that what creates the link between exchange and value is politics, construed broadly' (Appadurai 1986, p. 3). This particular view is built upon twin principles: that economic exchange creates value, and that value is embodied in commodities which are exchanged. Agreeing to these principles 'justifies the conceit that commodities, like persons, have social lives' (Appadurai 1986, p. 3).

These phrases capture an expression of power that is too often overlooked: that power accrues from the exchange of social values. While we readily accept that the exchange of ideologies is a precursor to power relations, the exchange of more discrete values of convenience and nutrition, for example, go unchallenged. Yet these are the very values that are embedded in the chicken by social actors, creating both economic and symbolic exchange possibilities for the meat relative to other foods. Those participating in the exchange of the chicken meat — the buyers, sellers and intermediaries — are negotiating not only the social and economic worth of the chicken, but their own relative status and power as well.

Both Kopytoff and Appadurai suggest that one way of charting cultural dynamism is to pay particular attention to the process of commodification, or commoditisation, to use their term. They describe the process as consisting of three elements. The first, the commodity phase, refers to the status of the good or service as it moves in and out of commodification. The second is the commodity candidacy, or the standards and criteria that define the exchangeability of items, while the third element concerns the commodity context, referring to the social settings of the exchange (Appadurai 1986, p. 15).

In relation to food, the first element incorporates the moving of production from the backyard and kitchen onto the industrial farm and into the factory. It is the realm of material production, including shifts from self-production to wage labour production, and from the household to the market economy. The second and third elements involve cultural production processes, including judgements of goodness, price, risk and trust within socially constructed contexts such as the family, public eating places and private households.

Appadurai adds a further concept that is useful for exploring commodity candidacy. He describes the 'regime of value' as 'a broad set of agreements concerning what is desirable, what a reasonable 'exchange of sacrifices' comprises, and who is permitted to exercise what kind of effective demand in what circumstances' (Appadurai 1986, p. 57). This particular heuristic device reminds researchers of two matters: the status of a commodity is a judgement made by people not by markets, and that a commodity has the ability to move in and out of status. It also alerts researchers to the idea that commodities may embody a hierarchy of values and that commodity status cannot be taken for granted, because the values attached to it reflect how it is embedded within social life.

Finally, as mentioned above, how one positions one's self in regard to any specific regime of value is fraught with uncertainty. Frow emphasises the process of receiving training in the assessment of value and nominates the institutions of mass education and mass media as playing decisive roles, detached as they are 'from local cultures and local communities' (Frow 1995, p. 155). His point that values are constructed through taken-for-granted institutions is a further principle worthy of guiding a study which must attend to the structuration of possibilities, not simply structural arrangements.[5] And as we see in Chapter 3, those most prosaic of institutions, the long-established supermarket chains, offer another setting in which we learn how and what to value.

From these works, I have taken seriously the challenge that to understand social change one has to understand the complex interactions between producers, consumers and the intermediaries located in the spheres of distribution and exchange. I have found only one study that has attempted to methodically traverse production, distribution and consumption and that is an account of the microwave. Cockburn and Ormrod (1993) follow the microwave's 'life trajectory' from the design office to the factory, the shops and into the household. Their study was similarly an examination of power, but rather than focussing on power relations between producers and consumers they focussed on the ways in which technologies are shaped by, and shape, gender relations. Commenting on their adoption of a multi-method, qualitative approach that used interconnecting case studies, the two researchers acknowledged that they risked generating a narrative that was thinly spread (Cockburn & Ormrod 1993, p. 4). Telling the contemporary story of the social life of the chicken runs a similar risk.

HOW THE BOOK IS ORGANISED

This first chapter is intended to provide a life and a structure to the very complex issue of culinary cultural change and to explain how *the social life* of one commodity, the table chicken, will be used to explore power in the Australian food system. Chapter 2 amplifies the debate begun in this chapter regarding the nature of change in culinary cultures. While there is little agreement in the literature on who influences whom, a recurring theme concerns the way in which production and consumption are linked at the level of ideas. Food systems operate relatively smoothly without producers and consumers *knowing* one another, but there is constant activity to forge a common language of concerns and aspirations about food and its place in the broader social domain. For this reason, Chapter 3 identifies those intermediaries who broker common ground, and focuses on the claim that retail traders are pivotal to this process.

Following the principles articulated above concerning the social life of commodities, insights from the literature on consumption, agricultural sociology and retail geography are used in Chapter 3 to develop a framework for collecting a mass of data on the table chicken. Chapters 4 to 7 contain material collected during the fieldwork, organised according to the modified framework. Chapters 4 and 5 confirm that neither consumption nor production sphere activities account fully for this particular food's popularity while Chapters 6 and 7 support an emerging theory which positions retail capital as the driver of food system change. However, as the case study indicates, retail capital accounts for only part of the power of those in distribution and exchange. A more plausible explanation for these actors' influence concerns their cultural activities. A host of different actors operating as cultural producers coalesce alongside retail-based actors in what I term the 'food producer-consumer services sector'. This sector creates channels for moving cultural practices into the marketplace and in so doing provides a commodity context that may or may not be commodity friendly.

The final two chapters apply insights from the social life of the chicken to understand how the balance of power is exercised within food systems and how globalisation intersects with culinary cultures. Chapter 8 conceptualises the process of mediation between producers and consumers by examining the interpenetration of cultural and economic processes. I argue that the potential to influence the social life of commodities and the culinary culture currently lies in the capacity of some actors to move goods between the circuits of culture and capital or, put another way, to mobilise particular cultural economy processes. The concept of market mediated authority is highlighted, given the significance of authorisation as one such process. Some important social theorists have questioned whether cultural economic processes are the hallmark of late modernity but few have attempted to articulate what a cultural economy looks like in the present era. I argue that the growing esteem of the table chicken reveals its operation.

Chapter 9 concludes the case study by suggesting the existence of a two-way relationship between globalisation and chicken meat's esteem. The evidence presented shows that while global restructuring pressures have been largely resisted by Australian producers and processors, all the signs suggest that a global chicken is not far away. While chicken production is likely to remain a domestic industry in most nations, global consumption data indicates that chicken is sought by working people worldwide, for both material and cultural reasons. Particular understandings of health and convenience are being disseminated rapidly and it is conceivable that 'healthy convenience' will become a *Zeitgeist* of an era marked by feelings of anxiety and overwork by women in particular. Through playing such an important role in the culinary cultures of so many working families, it is conceivable that the table chicken is fundamental to feeding the current wave of global economic activity, just as sugar, grains and red meat have done in previous eras.

All told, this research aims to produce an interdisciplinary study of the powerful appetites shaping food systems. Ideally the table chicken's story will provide an approach for others who wish to make food the centre of the plate, while using the social sciences to aid digestion.

2 POWER IN THE CULINARY CULTURE

One looks in vain for any discussion of food and eating in the work of most of the classic sociologists.

(Mennell et al. 1992, p. 1)

In a comprehensive overview of the field of food sociology, the co-authors of *The Sociology of Food: Eating, Diet and Culture* note the dearth of interest by the 'deities in the sociological pantheon' in theorising food and social relations (Mennell et al. 1992, ch. 1). Food, it seems, was of interest to Marx, Durkheim and Weber as a manifestation of inequality and for its role in the development of social structures. In short, food and eating have been traditionally studied for the way they illustrated more *substantial* (and possibly more male) sociological concerns. The overview moves quickly to assert the emergence of a new sub-branch of sociology, the sociology of food and eating. Mennell and co-authors provide several reasons for the development, including the mass media coverage of hunger on the one hand, and eating disorders on the other; the professionalisation of nutrition and dietetics; and the study of the 'trivial' and everyday life within another sub-branch, the sociology of culture. 'This particular rise is probably bolstered by a social respectability afforded by a shift in analytic and empirical attention from the sociologies of industrialised production to those of industrialised consumption' (Mennell et al. 1992, p. 5).

The issue of *how* the sociologies of culture and of consumption bring a different view of power to that deployed by the grand old men of sociology is sidestepped in *The Sociology of Food*. Despite the fact that 'power' based in economic relations has been a fundamental concept in sociology for over a century, the handling of power within the recent sociology of food texts is highly ambivalent. The term is not found in the majority of their indexes, and where power is addressed it is either in the context of gender relations or a 'Foucauldian' reading; rendering power as the asymmetric relations between male breadwinners and female housewives or between 'experts' and 'subjects'.

The general line of reasoning, built upon Foucault's work, is that diet and food choice have become medicalised by health professionals to serve their own needs, to control women and to discipline the body (Crotty 1995; Lupton 1996; Santich 1995b). While dietary advice is commonly argued to be a means of social control, there is little discussion as to the interrelationship between these forms of social control and economic power.

Just as the sociology of culture is credited, in part, with the emergence of food sociology, I would argue that the current dominance of cultural studies in the social sciences and the obsession with the body encourages a narrow view of power in much that is being called food sociology. Indeed, one could paraphrase the quotation with which this chapter opens, by remarking that 'one looks in vain for any discussion of power in the sociology of food and eating'. This situation is the more remarkable given that general texts on social theory claim a fundamental shift away from a focus on production to consumption. The following claims are fairly typical of what one encounters: consumption 'integrates and manages society, for the consumer has displaced the producer as the center of social engagement' (Miller 1993, pp. 47–48); '[c]onsumer conduct has assumed the importance of work — the cognitive and moral focus of life' (Lash & Urry 1994, p. 296); and 'the stratification process is now focussed on possibilities for consumption rather than production relations' (Waters 1995, p. 92).

This chapter begins by exploring the justifications for these rather extravagant claims. An examination of the way in which producer power is discussed follows, with particular reference to chicken meat production. The material suggests that the balance of power in culinary cultures is a contest waged between consumers and producers over how to think about food, as much as it is about economic power.

POWERFUL CONSUMERS

Consumers are described as more powerful than in (often unspecified) times past for numerous reasons, and I have chosen to concentrate on three prevalent arguments:

- consumer impact on the market through the exercise of choice while shopping

- the greater ease of feeding the family given the loosening of societal expectations on women, coupled with the availability of 'convenience foods'
- the ability to enjoy food more readily given the dissolution of long-accepted rules around eating (referred to in Chapter 1 as gastroanomie and the decline of the ritual meal), and the self-reflexive operation of desire.

While sociologists of consumption and of culture by and large write about the arrival of the empowered consumer, a few are more cautious about the type and extent of power they believe can be exercised in consumption sphere activities. The constraints on the consumer that have been identified concern the tenuous authority of the family cook, the fluid authority of the consumer, and the often unsatisfying exercise of desire, especially in relation to food. This last constraint has been attributed to a psychobiological phenomenon called 'omnivore's paradox' (Fischler 1988). Tensions about what to eat and what to feed others brings to the surface a perennial question: how does one determine what foods are good to eat? This question has provided the basis of a long-standing debate in anthropology. I describe the debate briefly because pronouncing, and reflecting on, food's goodness is a fundamental activity undertaken by producers, consumers and numerous others involved in distributing and discussing food. My research shows that the pathway to chicken's perceived goodness has been multifaceted, with the end result critical to consumer esteem.

SHOPPING AND THE EXERCISE OF CHOICE

> To an important degree talking of 'the consumer' is
> merely to dignify the term 'shopper'.
>
> (Warde 1994, p. 66)

Consumption can be broken down, as is shown in the next chapter, into discrete facets. Shopping is one such, and this daily activity is being taken seriously because, as the editors of *The Shopping Experience* put it, shopping is 'a paradigmatic case illustrating the fundamental shift in the structuring principle of society from production to consumption' (Falk & Campbell 1997, p. 2). The nuanced and contradictory experiences of shopping are explored through an examination of shopping sites including department stores, shopping malls and supermarkets. Shopping is portrayed by the various contributors as a positive experience, especially for women.

Two chapters of *The Shopping Experience* are particularly emphatic on this last point. Campbell, a co-editor of the text, argues that the gendered activity of shopping is creating a future in which women will have more collective power than men. He bases this conclusion on research that examined gendered attitudes to shopping. Campbell's data showed that women's shopping is underpinned by postmodern want and desire, while men's is motivated by modernist need (Campbell 1997, p. 169). Women apparently go to some lengths to keep the valued activity to themselves by minimising and disparaging men's shopping activities. According to Campbell, as shopping becomes more of a recreational and expressive pursuit, women's approach to shopping will position them as central to social action. He concludes that women's power is based upon the satisfaction they derive from shopping and their influence on the market through their purchasing decisions (Campbell 1997, p. 175).

No piece gives a more glowing assessment of women's power through being a female shopper than the chapter by celebrated anthropologist, Mary Douglas. She begins by dignifying the female dominated shopping effort by rescuing the woman shopper from the stereotypes of being an economics and fashion driven actor. Douglas argues that power is achieved through the refusal to purchase goods and that the female shopper's 'rationality' is most obvious in this form of protest (Douglas 1997, p. 17). More importantly, the woman's exercise of choice is linked to the type of society in which she wants to live and, in so doing, women become the major actors in shaping cultural institutions, particularly those of a culinary and economic nature. 'Shopping is reactive, true, but at the same time it is positive. It is assertive, it announces allegiance. That is why it takes so much deliberation and so much time' (Douglas 1997, p. 18).

Similar themes resonate through an earlier edited collection titled *Acknowledging Consumption*. In his self-styled polemic, Daniel Miller claims that 'consumption has become the vanguard of history' (Miller 1995, p. 1). He bases this assertion largely on the fact that so many disciplines are treating the topic seriously, with seven being represented in the book. Once again women are depicted as the more powerful agents. The questioning of products and of corporate producers which followed from the 1960s counter-culture movements has combined with western affluence to create a housewife with considerable power. In Miller's opinion, the aggregate decisions of housewives, exercised while shopping, gives them greater influence over capitalism than the International Monetary Fund. It is in this sense that Miller calls the western housewife a 'global dictator' (Miller 1995, p. 9).

While much of Miller's argument is deliberately stereotypical and reifying, he highlights the complex interrelationship between production and consumption, and the ambiguous authority and power of women for food provisioning. According to Miller, exercising thrift and looking after the family on the basis of the male wage no longer represents the extent of women's work and influence: women work outside the home for wages and they vote out governments who do not keep food prices low. However, their insistence on cheap foods minimises food producer and retailer profitability, and this leads to pressures for corporate cost-cutting through depressing wages and jobs: the employment on offer to women. Women, as he puts it, are both the victims and beneficiaries of capitalism (Miller 1995, pp. 8–10).

A further body of work deals with the way individuals are empowered through their capacity to interpret and create meaning. In an overview of the sociology of consumption, Humphery identifies consumption:

> ... as a potential arena of personal empowerment, cultural subversion, and even political resistance ... The 'consumer' was positioned as active, rather than passive, as the 'producer' of usages and meanings that the marketplace may not have assigned to a particular commodity or consumer space, and which potentially undermined or evaded consumerist ideologies (Humphery 1998, pp. 7–8).

This tenor applies particularly to those writing about shopping malls and department stores (Morris 1988; Reekie 1993; Zola 1992). One example will suffice. In his summation of the pleasure generated in the 'cathedrals of consumption', otherwise known as shopping malls, Fiske alleges that:

> [t]he values of commodities can be transformed by the practices of their users, as can those of language, for as language can have no fixed reference point in a universal reality, neither can commodities have final values fixed in their materiality. The practices of the users of a system not only can exploit its potential, but can modify the system itself. In the practices of consumption the commodity system is exposed to the power of the consumer, for the power of the system is not just top-down, or center-outward, but always two-way, always a flux of conflicting power and resistances (Fiske 1991, p. 31).

The material summarised thus far conveys a picture of consumers as powerful marketplace actors. *They* force markets to respond to their demands and then interpret the market response in ways only they control. Moreover, *they* hold governments accountable for market pricing policies, especially in relation to food. Women, as a group, are considered to be both the beneficiaries of the market as well as responsible for the products and experiences it offers. In this way, women are pre-eminent cultural actors.

FREED FROM FEEDING THE FAMILY

Over the last twenty-five years, a body of work has been devoted to exposing the household labour process and associated family dynamics, mainly highlighting the onerousness and devaluation which accompanies housework. Of this, the superlative research also portrays the mixed emotions with which women regard this unpaid labour, especially food preparation and feeding the family. While studies continue to show that little has changed despite the protests of, and policy directions advocated by, the second wave of the women's movement, some sociologists believe the burden of feeding the family is lessening. Generally, two contributing factors are acknowledged: changed expectations that meals have to be cooked from scratch by the wife and mother, and the market availability of products as replacements for home cooking. What follows here is a brief description of how food provisioning was portrayed in the 1970s and 1980s, along with recent research that shows how this particular facet of consumption is changing.

In a landmark study, not only for the sociology of food but for gender studies, Oakley sought in 1974 to describe domestic work as a 'job like any other'. She examined housewives' work satisfaction, routines and supervision, and found a vast disparity between women's experience of cooking and media portrayals of happiness in the kitchen. Oakley argued that media coverage of cooking as a creative entity was 'designed to subtract it from the category of "work" and add it to the creative pleasure dimension. This treatment of cooking ... is a particularly clear demonstration of how the social denial of housework as work operates' (cited in Luxton 1980, p. 145).

Oakley's conclusions were broadly supported by research carried out in the United Kingdom. In a relatively large qualitative study of the experiences of food provisioning of 200 mothers with young children, living in an urban area of northern England, Charles and Kerr (1986) found that cooking a 'proper meal' each day 'ready for their husbands' return from work' was central to the women's food preparation routines. Furthermore, they observed that men's food preferences and women's quality of married life were inextricably linked, for 'women cook *for* men, they cook to please men and to show affection for men' (Charles & Kerr 1986, p. 60). Women could name their husband's food likes and dislikes but had to be pushed to name their own, while children's food preferences came a clear second to men's. The authors argue that the sexual division of labour within the family is a relation of power which is constantly reproduced at meal times: the refusal of food, and husband's sometimes violent reactions to particular foods, meant that the food in question did not appear again (Charles & Kerr 1986, p. 62). This, and earlier research conducted by Murcott (1982; 1986), suggests that women's activities are controlled by men's relationship to food and drink: their likes and dislikes, their work routines, the size of their pay packets and domestic harmony (see also Mennell et al. 1992, ch. 13).

Recent evidence suggests, however, that anxiety about providing meals in defined ways may be wavering. Barbara Santich, a nutrition historian, found that among forty-nine Australian-born women, living in a low income area of South Australia, one-third were unreservedly enthusiastic about cooking, an equal number positively disliked it and another third 'didn't mind'. In speculative voice Santich notes that '[w]omen who dislike cooking, who profess little interest in it, and who derive little satisfaction from cooking, may be less content to adopt the stereotypical nurturing role' (Santich 1995a, p. 11). Santich suggests that the rejection of societal norms regarding cooking stems partly from the impact of the women's movement and its critique of the centrality of household duties to women's identity.

This rejection of societal norms may equally be fuelled by women's greater value to families and to national economies as wage labourers (Goodman & Redclift 1991; Marsden & Little 1990). In other work, Gofton (1990) juxtaposes E.P. Thompson's (1968) application of 'time disciplines' to working class males, with what he calls 'time famines' currently experienced by housewives. Gofton reasons that time famines are the result of women being used to create flexibility in the economy. In this account, women's mass entry into paid employment has had a desirable effect, because it has loosened the stereotype of the wife and mother. As a result of changed expectations, Gofton suggests that:

> [u]nder the roles and relationships of the new division of labour, meals and food provisioning have become far less significant events. Far less rides on them: the mother's self esteem no longer hangs on the kinds of food she provides for her family while the husband's unquestioned hegemony, on the basis of his role as sole/main provider, no longer demands to be celebrated in the form of the meal (Gofton 1990, p. 92).

Gofton and Ness (1991) elaborated these tensions in a paper provocatively entitled 'Who killed the lazy housewife?'. Here they refer to 'the new role of women in the social order', a role forged through the twin trends of concern for health and the need for convenience foods. The 'death of the lazy housewife' was chosen to represent the evaporating expectations that have defined the role of housewife: daily shopping trips, few labour-saving instant dishes and no take away meals.

The research by Santich, Gofton and colleagues indicates a shift is occurring in relation to the division of labour at home. The market is portrayed as empowering women through providing products that offer an alternative to 'the chore of cooking'. Although as a participant in a transport seminar I attended put it, the same trends can best be summarised as 'freed from the stove to be chained to the wheel'.[1]

CHANGING FOOD PRACTICES AND THE SELF-REFLEXIVE OPERATION OF DESIRE

In Chapter 1, changing culinary cultures were identified through the concept of gastroanomie and through the observation that the ritual meal is being displaced by practices such as grazing. Both Gofton (1990) and Falk (1994) have invoked the metaphor of food passing from the social table to the body of the individual as another way of signalling culinary change. Falk attributes this particular trend to the economic and social relations embedded in capitalism. He argues that the new food practices and products have liberatory potential because, as individuals are forced to negotiate their way through the market, they can satisfy wants and act on desire.

Falk's theory of consumption is based on evidence of a shift in the hierarchy between two oralities: speaking and eating. He identifies the United States at the turn of the 20th century as being the home of the 'modern oralities', and of fostering the decline of the ritual meal taken in company and in silence. Since then, the meal has been displaced by individual acts of feeding and even when food is eaten in company, the act of eating is dominated by talking.

Evidence of changing food practices has prompted others to proselytise the power of consumers. In one such account, Australian consumers who eat outside the home are posited as subverting civilising processes. In a phenomenological account of 'dining out', Finkelstein argues that eating out is particularly significant for women: they can be seen in public and they can imbibe, as men have over a long period, the 'spectacle and experience' offered by bars, cafes and streets (Finkelstein 1989, p. 5). In this way dining out, like take away meals and barbecues presided over by men, becomes an avenue of relaxation for women and, in Finkelstein's terms, 'a source of incivility'.

Desire, in Falk's theory of consumption, provides a non-reducible link between structure and agency. And so it appears in another Australian study, where power in consumption is said to start in childhood. On the basis of focus groups conducted among young Australian men and women, Lupton (1996) notes that:

> ... the pattern of dietary preferences and habits laid down in childhood never completely disappears; it is always reacted to, consciously or otherwise. Thus, a food or dish may be desired because of custom or habit, lack of awareness of alternatives or a longing for the comfort and pleasure of the familiar (Lupton 1996, p. 67).

Asserting one's independence from parental authority and family ties is, Lupton contends, an important impetus for trying new foods. Personal agency occurs, it would seem, in relation to affirming memories or rebelling against them. Additional agency accrues through individual reflection on all manner of dietary advice, and once again desire mediates the messages. Lupton concludes that individuals:

> ... do not passively adopt discourses in relation to food and eating. On the contrary... they take up ascetic discourses of self-control at some times and in some contexts in their quest to achieve the 'civilized

> body' and resist or ignore them at other times in their desire to engage in the release offered by hedonism and sensual self-indulgence (Lupton 1996, p. 155).

Resistance to nutrition discourse is a case in point. In a social history of dietary reform centering on the nutrition field, Crotty (1995) is particularly persuasive in showing how women have been the targets of what she terms the 'good nutrition': that is, advice about what is scientifically and medically acceptable in food. Crotty argues that, over this century, nutritionists have taken the place of women in the kitchen as far as knowing what is and is not good to eat. However, in a chapter entitled 'The subversive nature of everyday life', Crotty also questions how successful these messages have actually been. It is there she argues that the idiosyncratic, the unexpected and the unpredictable aspects of everyday life are a problem for nutrition researchers, and are to be applauded because 'they represent, on a small scale, a resistance to the tyranny of expertise' (Crotty 1995, p. 91).

CONSTRAINTS TO CONSUMER POWER

For many scholars the latter part of the 20th century is characterised by individual consumer empowerment and by consumer demands dictating market directions. Other reading shows that the nature of consumer power is highly problematic.

THE AUTHORITY OF THE HOUSEWIFE AND FAMILY COOK IS CONSTANTLY COMPROMISED

Even Miller, the proponent of the concept of the housewife as global dictator, highlights the various contradictory states in which housewives find themselves. His caveats on women's power are worth summarising. He remarks for instance that housewifery 'is one of the least valorised, most lowly and most commonly denigrated practices of the modern world' (Miller 1995, p. 35). Moreover, the source of a housewife's desire is subservient to 'the moral economy of the home' where consumption is mediated by the dynamics of family life within the domestic sphere. He also observes that rather than exercising choice, housewives are forced to negotiate an absence of choice.

A further insight is usefully added to this list of contradictory tensions. While Gofton and Ness (1991) report that housewives have been freed from 'the lazy housewife' syndrome, Gofton (1990) argues elsewhere that they have acquired new concerns regarding food. These arise from a heightened awareness of what Turner (1984) terms 'biopolitics', where social action revolves around bodily concerns like sexuality, reproduction and health. Gofton suggests women's concerns with health arise from their reliance on others for the family's food. Both Gofton and market research conducted in Australia highlight women's special concerns for the diet-related health of their children (Mackay 1992). So while women are spending less time preparing and cooking food, the amount of time devoted to worrying about food has not decreased (Gofton 1990). What emerges from the research is that the traditional authority of the female family cook is threatened and that for some, this is not a cause for celebration.

Three poignant accounts carried out on different continents attest to the resistance that some women feel for market relations taking over their housework, interpreted by some as emotional work. Luxton's description of a North American rural community showed that many women were very ambivalent about not cooking and foregoing the associated 'expressions of ... worth and caring' (Luxton 1980, p. 117). So were the women interviewed for Counihan's (1988) ethnography of the changing power and status of women in urban Italy. Her respondents were open about the mixed emotions of self-denial, creativity and appreciation by others as a good cook. They describe the enormity of the struggle to do everything without help from their husbands and in light of expectations from their mothers. Counihan found that the women coped by trying 'to remain the principal administrators of home and family at the same time as they hold full-time wage-labor jobs. Because of time and identity conflict ... they can perform neither well ... they are losing their traditional domestic influence over family and children' (Counihan 1988, p. 51). Counihan concludes that the women 'clearly feel ambivalence and conflict about their declining role in food provisioning. They want to control their family's foods, but do not have time because they also want to work' (Counihan 1988, p. 58).

The theme is similar in a moving account of men's violence around the family meal in Britain. On the basis of her study, Whitehead concludes by saying '[i]n cooking it is women who are speaking, and what I have tried to show is that what is spoken and what is meant is complex, deep-rooted and often quite contradictory. It contains elements of servitude, of power, of affection, of imposed and willingly embraced altruism, of self-denial and cheerful giving' (Whitehead 1994, p. 128).

FLUID CONSUMER AUTHORITY

In the studies by Murcott, Luxton, and Charles and Kerr, it becomes clear that women consumers believe that they know where power in the food system lies. These respondents generally expressed considerable feelings of impotence about what foods they could purchase and where they could shop. A general tenor was that women 'as individuals are powerless when faced with the organised strength of the food industry. And the women we spoke to were far from happy about the situation ... [Furthermore] [s]ome women openly doubted the myth of consumer sovereignty ... especially if trying to base food purchasing around health concerns' (Charles & Kerr 1986, pp. 65–66).

It is just possible that those who are promoting the power of consumers do so because they have gazed for too long upon department stores and shopping malls, fashion and leisure. In the little that has been written about food shopping and supermarkets the tone is more measured. Humphery (1998), who has recently published a social history of Australian supermarkets, attempts to steer a course between celebrating consumerism's potential and acknowledging its oppressions. He argues that instead of accepting the position that people 'make do' (Michel de Certeau's concept) it is important to acknowledge that people may be reflecting upon, or remaining distanced from, what is on offer. Humphery highlights how supermarket shoppers are fully aware that 'the frustrations are constant but usually endured, and the pleasures and evasions are real, but of poor quality, and are understood as such' (Humphery 1998, p. 17). However, he points out that for even the most reflective shopper there is little alternative to what the market is offering and, with acceptance of market goods, the purchaser acquires other uninvited lifestyle accoutrements — ranging from self-service, to packaging and the need for private transport.

It is in this context that the idea of fluid consumer authority is useful. The concept acknowledges that consumers rework the meanings attached to goods, but it questions the durability and strength of such consumer activity (Abercrombie 1994). And while Keat accepts that 'the first person knowledge of desires' is valued for the way in which it operates 'as the "transgression" of previously accepted, and authoritatively "enforced" boundaries', he wonders what this means in terms of relative authority, given that the market is 'the most powerful transgressor of boundaries, the most active dissolver of meanings [and] the most radical challenger of social authority' (Keat 1994, p. 39). Both authors suggest that producers have more sustained authority because they actively promote the rules and resources by which what is thinkable and do-able takes place: they structure the possibilities to which consumers respond.

DESIRE, THE MARKET AND OMNIVORE'S PARADOX

There is no agreement within the sociology of consumption as to what constitutes desire, how it operates or what place it has in consumption activities. For many, though, it is central.

On the basis that foods are ingested and contain all sorts of biological risks, I would argue that consumption of food is more vexed than consumption associated with fashion, household items, cars and housing. Fischler's work is particularly persuasive in this regard. Fischler describes the circulation of food between the material and corporeal as significant for the culinary order, for self-identity and for social identity. He refers to 'the act of incorporation' to describe 'the action in which we send food across the frontier between the world and self, between "outside" and "inside" our body' (Fischler 1988, p. 279).

Following his identification of the condition of gastroanomie, Fischler argues that human beings suffer from a biological need for food variety, or omnivorousness. As he puts it '[o]mnivorousness first implies autonomy, freedom, adaptability. Unlike specialized eaters, an omnivore has the invaluable ability to thrive on a multitude of different foodstuffs and diets, and so to adapt to changes in its environment' (Fischler 1988, p. 277). He argues that man cannot live without a minimum variety, which entails juggling survival with eating a range of potentially risky foods. This gives rise to a situation called omnivore's paradox, in which:

> ... each act of incorporation implies not only a risk but also a chance and a hope — of becoming more what one is, or what one would like to be. Food makes the eater: it is therefore natural that the eater should try to make himself by eating. From this principle of the making of the eater by his food stems the vital necessity of identifying foods, again in both literal and figurative senses. This is the clear consequence of the principle of incorporation: if we do not know what we eat, how can we know what we are (Fischler 1988, pp. 281–282)?

The principle of incorporation implies that culinary cultures are the outcome of biological as well as social factors. For Fischler, the absence of rules around food combined with the anxieties created by being omnivores, has shifted the control of food matters from cooks and eaters to producers and retailers. The problematic nature of food identification is laid at the feet of 'the recent expansion of the agro-industry and industrialized food production' (Fischler 1988, p. 289).

The processes by which food anxieties and perceived risks are managed are also of concern. In company with several well-known anthropologists (Douglas and Levi-Strauss), Fischler has advanced the notion that food and eating practices are a grammar which provides the basis of social group formation, intergroup hierarchy, social distinction and an individual's place in the cosmos. It was on this basis that Levi-Strauss proposed in 1965 that food had to be *good to think* before it was good to eat. He arrived at this conclusion through his famous culinary triangle: the raw, the cooked and the rotten. By juxtaposing the three food states he arrived at formulae (called binary gustemes) for how nature is transformed, with consequences for how those same foods become esteemed, or good to think.

Harris (1986) was one of the first to disagree with the proposition that humans eat only after having judged the food's symbolic goodness. He rails against the idea that 'foodways are arbitrary' and believes all food choices can be explained by nutritional, ecological or economic reasons (Harris 1986, p. 15). He argues that meaning systems are created through eating and that the cultural sphere is the product of materialism. Much like Marx one hundred years earlier, Harris argues that agriculture precedes culture, and in debate with Levi-Strauss says that '[f]ood must nourish the collective stomach before it can feed the collective mind' (Harris 1986, p. 15).

The literature makes it hard to deny, however, that consumers' capacity to reflect on the foods they eat is increasing alongside their growing propensity to question a range of authorities, knowledge and experts on a host of matters. This coincides with a generalised desire to be self-improving, to be socially mobile and acquire what Bourdieu (1977) identifies as 'cultural capital'. The combined scepticism and self-reflexivity sets up what Lash and Urry (1994) describe as the pre-condition for the reflexive accumulation strategies of producers.

PRODUCER POWER

Under conditions of fluid consumer power, to what extent and in what ways do producers exercise power? In this part of the chapter I concentrate upon the influence of producers over food systems in general, chicken meat commodity complexes specifically, and over consumption sphere activities.[2]

The most coherent corpus of work explaining power in food systems is the new political economy of agriculture. Emanating from North America, with a substantial Australasian contribution, the work comprises critical analyses of commodity complexes, and national and local food systems. Five books summarise the current thinking of this school: *The Global Restructuring of Agro-Food Systems* (McMichael 1994), *From Columbus to ConAgra: The Globalization of Agriculture and Food* (Bonanno et al. 1994), *Globalising Food* (Goodman & Watts 1997), *Globalisation and Agri-Food Restructuring: Perspectives from the Australasia Region* (Burch et al. 1996) and *Restructuring Global and Regional Agricultures* (Burch et al. 1999). Special editions of four journals, *Political Geography* (1993), the *Review of International Political Economy* (1994), *Nature and Society* (1997) and *Rural Sociology* (1999) are also devoted to research in this area.

For a decade or more, this relatively large group of academics has been arguing that no matter where in the world the farmer or peasant lives, the terms of their livelihood have been significantly altered by the strategies of transnational food corporations and the regulatory policies of states. They allege, too, that since the end of the Second World War state policies have been increasingly influenced by supra-national organisations, such as the International Monetary Fund and the World Bank, and multi-lateral agreements, such as GATT and the Uruguay Round. These scholars are unified in arguing that the dominant force restructuring agricultural relations across the globe is transnational capital, and that this form of capital is undermining the social contract between states and agricultural sectors (Ufkes 1993). Only rarely is the role of consumers addressed.

Given the volume of the political economy accounts of agriculture, I have selected three focal points from which to explore producer power. The first revolves around the transition from one food regime to another, especially as reflected in the shift away from the backyard, family-based extensive system of farming to the global and corporate, intensive system of agriculture. The food regimes theory, referred to in Chapter 1, describes specific interrelationships between production systems, regulatory regimes and capital accumulation systems. The negotiation of regulatory regimes and the adoption of particular capital accumulation strategies raises questions about the nature of power wielded by agrifood producers. Just as the power of consumers was shown in the previous section to be contingent on numerous factors, so economic geographers have been revealing the constraints on productive capital and corporate power. I describe how economic geography factors combine with features of the system of regulation to create 'uneven development'. An economic geography analysis of the Canadian poultry industry by Bowler (1994) is used to illustrate this argument.

The second point takes up the suggestion that we are witnessing, if not a third food regime, a system of flexible accumulation based upon flexible production systems. According to some, poultry complexes are leading the way in this form of accumulation. In this most recent system of capital accumulation, consumers are alleged to have unprecedented influence over how the market in agrifoods is organised, and what products are on sale. Questions are raised, however, as to whether chicken meat complexes everywhere exhibit flexible production characteristics. Chapter 5 addresses this issue relative to the Australian market.

Finally, the question of food's uniqueness, identified by sociologists and anthropologists and canvassed in the preceding section, raises potential challenges and opportunities for producers as well as consumers. Owing to the mass volume, low profit margin nature of much of the agrifood industrial sector, the strategy of product differentiation is vital to food company profitability. This particular strategy is discussed for the way it operates in a dialectical relationship with consumers exercising power through their own assignment of values to commodities.

FROM EXTENSIVE TO INTENSIVE ACCUMULATION TO UNEVEN DEVELOPMENT: THE CASE OF CHICKEN COMMODITY COMPLEXES

In their discussion of the coping strategies being adopted by Australian farmers in the face of agricultural restructuring, Gray and co-authors (1993) rely on the so-called 'French regulation school' to explain what forces lie behind the current restructuring. Indeed, regulation theory, which was developed originally to explain changes in the manufacturing and service sectors, is being widely applied in rural sociology to explain changes to rural society (Bowler 1994, p. 347; Ufkes 1993, p. 194). Several of the theory's key concepts are outlined below.

According to Bowler, regulation theory describes capitalism's passage through a series of regimes of accumulation, with each regime involving 'a relatively stable, coherent, temporal phase of economic development'. Each phase is identified in turn by a dominant system of accumulation, this system being defined as comprising 'an interrelationship between production and consumption over an extended period of time' (Bowler 1994, p. 347). A further vital ingredient to a regime of accumulation is the mode of social regulation, this being a set of institutional arrangements governing the wage relation and the form of state intervention.

The most common regulatory regimes schema is that identified by the terms Fordism and post-Fordism. Fordism was preceded by several forms of an extensive system of accumulation in which individual workers and entire families were enticed into factories. Once workers were assembled under one roof, a century of innovation brought increasingly detailed control of labour and mechanisation. Fordism refers to the period of the 1920s onwards, in which an intensive system of accumulation grew around the mass production of standard products and high wages to a low skilled workforce and later a welfare state to ensure a market for the goods produced.[3] Homogeneity and standardisation in both the production techniques and consumer preferences was critical to Fordism's success.

With the help of regulation theory, Friedmann and McMichael developed a food-based theory of geo-politics covering the last 125 years (Friedmann 1990; Friedmann & McMichael 1989; McMichael 1996). The theory involves two food regimes that correspond with Fordism and its immediate antecedent, and receives substantial support in a detailed account of the development of Australian agriculture (Lawrence 1987, ch. 4). The authors argue that two foods, grains and meat, have underpinned the successive food regimes by supplying the diets of working classes and by providing a source of capital through which corporations could build complexes which they controlled. The operation of the two food regimes and associated regulatory regimes has also been used to explain the evolution of poultry complexes in North America.

Friedmann (1990), for example, has uncovered the importance of poultry and egg production to the meat and grains commodity complexes over the last century through her detailed research of the history of wheat farms in the United States. In essence, the wheat farm was a mixed farm from where several commodities were produced and sold, some of which were the result of women's unwaged but commercial activities. At a superficial level, the farm women's value lay in their unpaid food production activity and their feeding of the farm family and seasonal workforce. At another level, the *farmer's wife* produced foods that could be bartered and sold, including eggs, poultry, cream and butter. The mix of commercial and unpaid labour undertaken by women on wheat farms carried the farm over bad seasons and made them viable over the long term. However, Friedmann found that this activity was neither officially nor privately valued. In the case of the United States:

> ... men on farms and in government derided poultry production as part of their devaluation of women's contribution to farm work. The invisible work of women was hidden in the household, they kept their egg money separate, reinvesting it in poultry production, but also bartering eggs with local merchants in return for goods needed for the household, for herself, and for the children (Friedmann 1990, p. 200).[4]

The free-ranging poultry moved 'inside the barn' during the Second World War for several reasons. One was the increased demand for eggs, another was a need for jobs for returning servicemen, and a third concerned making use of the grain surplus that the United States had generated. Friedmann and McMichael explain the shift in attitude to poultry production in the following way:

> During World War 2, with the encouragement of the US state, the breeding and rearing of livestock, especially poultry, was transformed (with varying degrees of success and speed) from handicraft and extensive techniques of husbandry to intensive, scientifically managed continuous production systems. From the input side, increasingly concentrated livestock producers required feedstocks and many other goods and services. Manufactured feeds were purchased from corporate manufacturers which sold only mixes of protein and caloric and other ingredients, a practice which originated in order to evade the price scissors created by US price supports for soybeans and price maxima set for soycakes by the American government as part of a policy to intensively restructure wartime meat (especially poultry) production (Friedmann & McMichael 1989, p. 106).

These accounts of poultry production's transition from an extensive to intensive system, using first women's unpaid labour and then men's paid labour, accords with Bowler's study of events in Canada. Bowler describes how poultry production until the 1930s was subject to a period of extensive accumulation, one in which poultry ranged free on hundreds of farms, fed on grains sourced from the farm or local grain merchants. Bowler identifies 'two sub-periods of restructuring' as changing the family farm basis of production under the Fordist regime of intensive accumulation. Firstly, specialised poultry producers were encouraged to move birds into 'deep litter' sheds in which their whole lives could be managed on the basis of new feed supplies and new feeding machinery, the result being the marginalisation of 'many thousands of small poultry producers' (Bowler 1994, p. 348). And from the 1960s onwards, more intensive production systems were introduced into poultry production whereby:

> ... 'factory farming' was employed to describe this capital-intensive system of agriculture which continued to favour the large specialized production unit at the expense of the small ... Moreover, poultry farms exhibited several characteristics of Fordist production, namely specialization of the labour function, mass production of standardized goods, factory-like premises, assembly line production techniques, and supply for the mass consumer markets; producers were able to exercise control over the biological process so as to co-ordinate production and labour time and remove their dependency on the land-base (Bowler 1994, pp. 348–349).

Chapter 5 reveals that Australia's poultry industry was subject to similar production and capital accumulation regimes. It also shows how the Australian poultry industry was transformed from hundreds of farming families to a handful of corporations that could control the genetic, feed and labour inputs necessary to produce chicken for a mass market.

Much of the material describing worldwide transitions in agriculture since the 1970s portrays a one-sided contest, with the largest corporations having an almost unfettered ride to dominance. However, economic geographers are reminding researchers of the importance of fixed as well as socially constructed geographies to agricultural systems. They show that national and regional physical peculiarities mediate the intentions of transnational and national actors alike. Nature, and the history of dealing with it, is central to their concerns. For instance, in a study of Australia's dairy and vegetable industries, Pritchard (1995; 1996) describes how global capital is forced to engage differentially with different sectors of agriculture. In the case of the dairy industry, he found that global capital has had an indirect effect only and that national capital continues to play the pivotal role. In contrast, within the vegetable processing sector, transnational corporations (TNCs) are consolidating their position through takeovers of Australian companies. Pritchard concludes that his case studies illustrate how sunk costs and human agency help to determine the geographical shape of agrifood globalisation. His work is important for the way it highlights that:

> [a]s global capital transcends national boundaries in search of profit opportunities, its behaviour is influenced by the landscape it confronts. Whilst the shift to globalisation may be a prevailing feature of recent agri-foods restructuring, its concrete manifestations are necessarily complex and diverse (Pritchard 1995, p. 50).

Similarly, Grant (1993) emphasises how the particularities of a history of firm decisions, government policies and cultural understandings of agriculture's place in the national psyche are important in terms of future investment decisions. Indeed, a major finding of Bowler's research on the Ontario poultry industry is the uneven development to which it is subject. Uneven development, in his opinion, is particularly pronounced in agriculture because of its land-based nature and the exploitation of biological processes (Bowler 1994, p. 348).

FLEXIBLE ACCUMULATION AND A POSSIBLE THIRD FOOD REGIME

In some people's opinion, a flexible system of accumulation is superseding Fordist or intensive accumulation, with the chicken meat complex showing the way. This third phase in capitalism's trajectory is called post-Fordism, and is underpinned by a mode of state deregulation and by flexible specialisation, or 'small-batch production of a variety of products, the use of flexible machinery and micro-electronics, and the employment of a skilled and flexible workforce' (Probert 1994, p. 101). A flexible workforce requires what Mathews (1994) terms functional and numerical flexibility — the former describes the multiskilling of workers, and the latter refers to achieving labour flexibility through part-time and temporary work, outsourcing and subcontracting. A number of forces are said to be responsible for Fordism losing ground: market saturation, greater differentiation in consumer preferences, consumer concerns for quality and not just quantity, speedier development of product innovations, and new distribution systems, such as Just In Time.[5] Flexible specialisation harks back to the craft production that dominated early forms of industrialisation and is currently valued worldwide because it provides a source of permanent innovation (Piore & Sabel 1984, p. 17).

Within the new political economy of agriculture, Friedmann's work is notable for dealing explicitly with the interrelationship between regimes of production and consumption. She proposes that each regime of accumulation is accompanied by distinctive norms of consumption. 'These norms refer not only to how much people use ... but also how they acquire goods, especially the extent and character of market dependence' (Friedmann 1990, p. 197). According to Friedmann, the shift from unwaged self-provisioning work to paid work represents not additional goods, but different goods, whose sale represents new value for capitalists. In a clear reference to the interdependence between consumption norms and regimes of accumulation, she argues that:

> [t]he key to the shift from extensive to intensive accumulation in the US (and eventually in other advanced capitalist economies) was the penetration of commodity relations deeply into daily life, transforming

> all aspects of self-provisioning by individuals, households, families and communities into commodity relations involving waged labor and purchased products (Friedmann 1990, pp. 197–198).

In *From Columbus to ConAgra,* Bonanno and co-authors believe the demands of particular consumers are behind the new production systems. 'As flexible has become the watchword for the global producers, flexible consumption — the recognition of the existence of a multiplicity of niches — has become the critical element in the globalization of consumption' (Bonanno et al. 1994, p. 11). The book claims that new production systems, characterised by Just In Time sourcing of inputs, global sourcing of inputs and flexible labour markets are catering particularly to global city inhabitants who are cosmopolitan in outlook. Furthermore, the new cosmopolitans have distinctive norms of consumption, including particular diets based upon what have been termed elsewhere as high value foods (HVFs). These foods are distinguished from traditional export commodities by being fresh rather than durable (Friedland 1994; Watts & Goodman 1997). HVFs, which include poultry, fruit, vegetables, shell fish and dairy products, allegedly contain post-Fordist attributes such as heterogeneity, 'quality' and a basis in market niches. These features 'place considerable weight on the point of consumption insofar as HVFs have to be culturally constituted for particular sorts of taste, diet, and "vanity"' (Watts & Goodman 1997, p. 11).

Not only are poultry complexes portrayed as early leaders of intensive production systems in agriculture, they are heralded as leaders in flexible accumulation strategies, or strategies that are responsive to consumer demand (Boyd & Watts 1997). In a study of the American industry, the point is made that no 'other agricultural commodity or agro-industry can match the capacity of the firms in the broiler industry who adjust production and develop new products with astonishing speed and flexibility …' (Boyd & Watts 1997, p. 215). The researchers note too, the extent of Just In Time provision of inputs into broiler production and the need for producers to be highly responsive to retailer demands. They conclude that particular territorial complexes in the south of the United States 'resemble (at a surprisingly early historical point) the flexible, just-in-time production systems customarily associated with the "new industrial districts" (the Third Italy) and the Japanese manufacturing revolution ("Toyotaism") of the 1960s' (Boyd & Watts 1997, p. 206).

Making use of Friedmann's work, two Australian agrifood academics support the view that the poultry industry exemplifies flexible specialisation and the tenuous beginnings of a third food regime. Friedmann has made the point that '[w]hile privileged consumers eat free-range chickens prepared through handicraft methods in food shops, restaurants, or by domestic servants, mass consumers eat reconstituted chicken foods from supermarket freezers or fast food restaurants' (Friedmann 1991, p. 86). Lawrence and Vanclay, in research on emerging niche markets for beef, extrapolate from this observation to suggest that:

> [a]lthough standardized and highly processed foods remain a key element in global food distribution, the metropolitan nations are experiencing — as part of the crisis of Fordism — rejection of the very techniques, methods and products that so successfully tied food production to consumption in the postwar years (Lawrence & Vanclay 1994, p. 93).

They suggest that niche markets and flexibly produced agrifoods may constitute a successor regime to the Fordist production approach of the second food regime. Chicken provides, in their estimation, an exemplar of an unfolding third food regime.

These views are not universally held, however, and Kim and Curry are more circumspect about labelling the American broiler production system post-Fordist. They found product differentiation to be 'as much [about] the marketing techniques, such as store design and advertising, that create the variety, as it is the actual ingredients of the food' (Kim & Curry 1993, p. 75). While they acknowledge an enormous range of chicken products to suit every taste, they assert that as 'the technical means to manufacture efficiently differentiated products are developed (that is, flexible production systems) they are employed, not, as according to Piore and Sabel (1984) to re-invent craft production, but to mass produce variety' (Kim & Curry 1993, p. 74).

The existence or otherwise of flexible production systems and of new norms of consumption permeates the discussion of consumer and producer power throughout the book and I return to the presence or otherwise of a new food regime in Chapter 9.

THE CONTINGENT PRODUCER AND THE UNIQUENESS OF FOOD

The power of producers appears from the foregoing to be constrained by several factors: norms of consumption, economic geography features and modes of regulation that influence the flow of capital. Such constraints apply to all commodity production, but even in political economy accounts one finds references to the uniqueness of food commodities *vis a vis* other commodities (Fine 1994; Goodman & Redclift 1991; Murdoch 1994). Food has three notable features: a perishable nature; finite demand at the level of the individual, even in conditions of relative abundance; and immediate material as well as psychic risks and benefits, as discussed earlier. The first feature compounds the second and third. In order to make durable what is inherently perishable, the firm as well as the individual consumer entertains risks: both financial and microbiological. In this section I address how these features shape production sphere activities.

In order to realise profits in the 20th century, food corporations have adopted two principal strategies. The first requires finding new markets and trading food internationally by moving into processed or durable foods (a feature of the first and second food regimes). The second involves value-adding for the domestic market, where greater profits per item are to be made: a feature of the second food regime, and the likely basis of any successor regime (Lawrence 1987; Leopold 1985; McMichael 1994). Most agree that at a certain point of affluence, national food systems are characterised by a particular norm of consumption: any 'extra expenditure by consumers tends to be used for the purchase, not of more food, but of more expensive food ...' (Lawrence 1987, p. 112). This in part explains the extraordinary lengths that food firms go to process or add value to foods, especially when the food market is saturated by new food products every year.

The strategies of finding new markets and value-adding are illustrated in Leopold's study of how agribusiness has evolved as an industry sector in the United States. Referring to the period immediately after World War 2, Leopold points out that agrifood businesses have had to address a general problem for capitalism, that is, 'the downward trend of the rate of profit' (Leopold 1985, p. 316). However, food companies have also realised that whilst profits from food processing may be slim, food processing is recession-proof because everyone has to eat. Leopold argues that food companies have charted their strategic direction within the scissors of this dilemma for forty years, a period in which they have invested heavily in product differentiation. According to Leopold, product differentiation is a precursor to industry level concentration and oligopoly power. Both are achieved through massive recourse to advertising, a cost which most small firms cannot bear. McMichael (1993) similarly argues that value-adding becomes essential in an inelastic market, and that the capacity to add value is accompanied by concentration. This capacity provides one explanation for the growth in transnational food corporations as well as the diminishing terms of trade for farmers.

Limited capacity for infinite food consumption in an era of food over-production motivates food producers and manufacturers to promote their product as *good to think*. As a result, food producers associate their products with health claims and try to influence national dietary guidelines so that the guidelines include what is already on offer in the marketplace. In reference to the United States, but applicable to Australia, Nestle writes:

> [i]n 1990, the US food supply provided an average of 3700kcal per day for every man, woman, and child in the country. Most adults need one half to two thirds of that amount, and most children even less. Overproduction means that any choice of one food product necessarily implies rejection of another ... Any suggestion to reduce intake of

a food component for reasons of health threatens the competitive advantage of any product containing that component (Nestle 1995, p. 273).

An examination of Australian chicken meat product development, product differentiation, marketing and advertising yields a complex picture of the interrelationship between producers and retailers to associate their goods with health and convenience in an attempt to get consumers to eat more chicken. However, it also reveals that corporations do not simply produce and trade in foods and their symbolic worth. Foods also promote corporate identities, and it is here that we learn the importance of chicken meat to the market place positioning strategies of corporations.

CONCLUSION

The foregoing literature review explores the controversies and starkly contrasting assumptions embraced by different disciplinary and theoretical perspectives. Even within the sociology of consumption there are opposing views of consumer sovereignty, and of women's power in particular. Western women shoppers and housewives are, at one extreme, being called the most powerful global actors of the moment, exercising leverage over governments and cultures. Other work, however, finds that not only is family food provisioning a set of activities riddled with tensions, the act of eating appears to be replete with contradictory promises. The much-heralded era of body politics is delivering only limited fulfilment and emancipation. Furthermore, consumer power in relation to food is heavily contingent on the operations of the market, or anonymous others, to provide the bulk of food. This is cause for both comfort and ambivalence in a context of omnivore's paradox and a desire to continue to be involved with food provisioning.

Even though many food sociology accounts barely mention food producers outside the home, production sphere activities *in toto* are accorded a significant role in shaping what happens in consumption. However, it is within the new political economy of agriculture that we learn of producer power's state of flux due to the restructuring of regulatory arrangements between states and agricultural sectors. Producer power is, at the very least, shaped by regulatory regimes, previous investment decisions and the natural environment. These global and local factors create the conditions for uneven development, a feature that is particularly pronounced for agrifood commodities due to their *natural* origins. Food's uniqueness opens up threats and opportunities for producers and consumers alike, giving rise to product differentiation, on the one hand, and omnivore's paradox on the other.

With product differentiation strategies assuming greater importance for profitability, the value-adding activity of producers has become more onerous. Studies of the North American poultry complex arrive at opposing conclusions regarding whether product differentiation strategies are responsive to consumers or whether they are simply the producers' means of maximising diminishing returns on investment. Whatever the conclusion on this matter, the balance of power between producers and consumers appears to be a struggle over who can define what is good to think in relation to food. And in a context of fluid authority relations, those who can mobilise the attributes necessary to be considered authoritative are well placed to exercise the balance of power.

3
CONSTRUCTING THE SOCIAL LIFE OF THE CHICKEN

The previous chapter reviews the sociological literature for explanations of the power that is exercised by producers and consumers. It reveals that most food-related research ignores the dynamics between production and consumption, with power being analysed from within, rather than between, the spheres of production and consumption. The struggles between producers and governments dominate in the new political economy of agriculture, and contestation over meanings, identities and interpersonal power underscores the sociology of consumption. Table 3.1 depicts the current state of knowledge about production and consumption.

Table 3.1
Common depictions of power in food production and consumption

PRODUCTION	CONSUMPTION
Economy	Culture
Structure	Action
Wage relation	Commodity relation
Instrumental body	Subjective body
Cognitive rationality, supply/resources/availability	Intersubjectivity, desire, demand/needs/acceptability
Public realm: workplace/state	Private realm: household/family
Exchange value	Use value
Men	Women

It is not surprising, therefore, to note the emergence of critiques on the dominant approaches to studying power in commodity complexes and food systems. For instance, Fine and Leopold (1993) admonish sociologists for treating consumption as a reflection of production. Mintz (1996), the author of a celebrated social history of sugar, also insists that greater attention needs to be paid to the processes of change which jointly engage both producers and consumers. In a similar vein, Arce and Marsden point out that 'food production and consumption are highly contingent and are reliant on delicately balanced alliances and social and economic arrangements' (Arce & Marsden 1993, p. 295). They specifically accuse those adopting a structuralist perspective, and they include regulation theory here, of overlooking 'the importance of actors' cultural and knowledge negotiations in defining the meaning of food'. They add:

> ... the political economy perspective on the food system has reached its empirical and conceptual limits ... Originating in an evolutionary analysis based on a Wallersteinian perspective on the world ... it has become somewhat limited in defining and interpreting social and spatial diversity (Arce & Marsden 1993, p. 296).[1]

Furthermore, much of the analysis shows a degree of perversity in relation to women's power. Assumptions and sweeping generalisations dominate the accounts of women's work and what constitutes that work. The class-based accounts all too often ignore gender and vice versa. Invariably, women's domestic labour, including food processing and distribution, is not included in national accounts figures even though the annual value of meals produced in Australian homes is estimated to exceed the value of goods produced in all of the country's manufacturing industries (Ironmonger 1989). As Waring (1988) puts it, these activities are outside the 'productive boundary' and as such 'count for nothing'. Seccombe similarly highlights that from a production perspective, the household constitutes a 'hidden abode' preventing us from being aware of 'the (centuries old) decline of various forms of commodity production in and around the household' (Seccombe 1986, p. 56). This means that paid and unpaid work are rarely considered to be symbiotic, with research by Friedmann (1990) and Goodman and Redclift (1991) being exceptional. How food systems and culinary cultures operate on a daily basis remains obscure as a consequence, making it difficult to reach an understanding about how power within the culinary culture is both reproduced and transformed.

Acknowledging the criticisms and being alert to the principles underlying the social life of a commodity as outlined in Chapter 1, I sought to implement a framework which would traverse the spheres of production and consumption and that would include the range of actors responsible for shaping power relations. No one framework was sufficient. What follows is a description of the modifications that were made to two frameworks: one that covers production sphere activities and another which encompasses the process of consumption. Neither framework, however, crosses adequately the binary divide depicted in Table 3.1. With this in mind, it became important to elaborate the actors and processes critical to the sphere of distribution and exchange. This chapter includes material that allows that elaboration to take place, leading to a comprehensive framework for examining the social life of the chicken.

THE STUDY OF PRODUCTION AND CONSUMPTION

In 1984, Friedland proposed a Commodity Systems Analysis (CSA) framework for describing the stages through which a commodity is transformed and acquires value. He challenged us to think of commodities as entities with a social as well as a physical presence some years before the authors quoted in the preceding section. He gave commodities a social life by reminding researchers that people's labour and ideas, their technological developments, the power circulating between groups, the way individuals co-operate, and their organisational structures, are all critical 'inputs'.

The CSA model is based on the recognition that agriculture has shifted from mixed farming and self-provisioning to single commodity-based farming and market-based consumption. The schema was developed on the basis of a number of agrifood commodity case studies conducted in California from the mid-1970s onwards, and provides an analytic process 'to recognize when and where interpenetration of systems occur, where the system being analysed touches upon other systems or is significantly affected by others' (Friedland 1984, p. 223). The model brings neo-Marxist and agricultural and industrial sociology insights to bear upon the problems being experienced by family farms, rural communities and agricultural labour.

Friedland's model was chosen as a starting point in the present study for three reasons. It forces the disciplined organisation of a mass of data and as a result lends itself to comparative analyses of commodities. Indeed its clarity of purpose makes it ideal for adaptation. Secondly, its premise that commodity systems have a 'social reality' encourages an approach that demands an actor orientation that is context and case specific. Finally, it was designed to identify the processes underlying the balance of power within a food commodity system.

According to Friedland's early work, all commodities are said to be the outcome of the following processes:

- *production practices*: includes production techniques, commodity characteristics like diseases, production cycles and associated problems

- *grower organisation and organisations*: farmers or growers are acknowledged as the basis of agriculture in this model: how they use their labour, whether they manage the labour of others, how they organise and how their organisations relate to others relevant to the commodity

- *labour as a factor of production*: includes the labor process, the institutionalisation of employment practices, and the presence and activities of unions

- *science production and application*: involves scrutiny of the knowledge base behind the productive activity, the links between research and development units and growers' groups, funding sources and the degree of public sector involvement

- *marketing and distribution networks*: involves the marketing of commodities and price setting arrangements.

Friedland encouraged researchers to refine the model and I have added two processes. Guided by the new political economy of agriculture, which emphasises the regulatory specifics of commodity complexes, it seems necessary to add a *regulatory politics* process to the production sphere, the focus being state-producer relationships.[2] Moreover, given the quantity of raw food product that is packaged and/or combined with other ingredients, it is timely to include a post-harvest process called *product design*, as distinct from the science production which occurs pre-harvest. Design has become an especially important part of the value-adding process given the specific limits to food consumption and the necessity to be permanently innovative under conditions of flexible accumulation.

While Friedland's argument enriched the concept of commodity production, it remains essentially a productivist perspective. Since the publication of Friedland's original framework, as Chapter 2 shows, entire sub-disciplines have devoted themselves to examining how power is shifting from producers to consumers. Furthermore, some are claiming that, contrary to much traditional Marxist thinking, individuals find meaning in consumption and not through their productive capacities. Indeed, several Marxist inspired consumption theorists have surfaced. Both Fine (Fine 1994; Fine, Heasman & Wright 1996; Fine & Leopold 1993) and Warde (1992; 1994; 1997), provide sustained rationales for why it is important to study consumption separate from, but mutually constitutive of, production. They propose that each experience of consumption is unique because each commodity is unique. Their respective systems of provision models allow greater agency on the part of individuals than do regulation theory based accounts, where norms of consumption reflect the intersection of particular production processes and modes of social regulation. Unlike Friedland, who concentrates upon production in the market place, both Fine and colleagues and Warde emphasise changes in the modes of production from household to market, and from reciprocity to market exchange.

Of the two provision models, Warde's is the neater and is used here to explicate the steps involved in consumption. Warde argues that each consumption 'episode' consists of a number of 'distinct facets': the process of production or provision; the conditions of access; the manner of delivery; and the environment or experience of enjoyment. In my opinion, this last facet necessitates highlighting: it is at one and the same time the setting for the consumption episode and the resulting experience. In the present study, it may be helpful to distinguish between the two, and indeed Warde (1992) has written at length about the experience of consuming.

The clarity of Warde's consumption processes is in keeping with the clarity of Friedland's production processes, but one major problem exists in bringing together the two frameworks. Warde's schema incorporates a range of producers: primary (farmers), secondary (the manufacturer or processor) and tertiary (the meal preparer). For the purposes of identifying shifts in activity between households and the market, the tertiary producer will be incorporated into the consumption process, and the primary and secondary producers into the production process. Given my desire to move away from categorising men as producers and women as consumers, this demarcation is far from satisfactory and a preferable approach would be to account for tertiary production in both production and consumption. While acknowledging this deficit, I am proposing that food consumption be conceived as having the following components:

- *tertiary production practices*: refers to who does the preparing — food service outlet, household member, workplace canteen or friend
- *means of access*: each of these modes is accompanied by 'market exchange, familial obligation, citizenship right and reciprocity' (Warde 1994, p. 19)
- *manner of delivery*: from help-yourself to being served
- *the eating environment or context*: refers not only to place and time but to social considerations, such as public or private eating and convivial or solitary eating
- *the eating experience*: involves the emotions accompanying what is consumed, ranging from pleasure to ambivalence to disgust, as well as self-reflexive activity in relation to identity.

EXTENDING THE COMMODITY ANALYSIS FRAMEWORK INTO DISTRIBUTION AND EXCHANGE

In the opening chapter I proposed that the way in which cultural and material production align is fundamental to shaping food tastes and to the dynamism existing in culinary cultures and food systems. To this end, actors must work hard to shape both the status of the commodity and the context in which commodities are exchanged. This section describes what these particular shaping processes look like. In the original version of the CSA model, Friedland (1984) points out that primary producers were often captured, or subsumed, by the distribution process. He did not, however, expand upon the process by incorporating its actors and dynamics, as I am suggesting. And while Marx argues against the artificial separation of the spheres that constitute the political economy, he does ascribe production, distribution and consumption with specific properties. He refers to distribution activities as 'special', and in *The Grundrisse* says distribution 'determines what proportion (quantity) of the products the individual is to receive; exchange determines the products in which the individual desires to receive his share allotted to him by distribution' (cited in McLellan 1973, pp. 32–33). Elaborating the processes of distribution and exchange is justified, in my opinion, by the evidence emerging from retail geography and cultural studies that distribution and exchange sphere activities are fundamental to food's accessibility, both metaphorically and physically.

When approaching the transformations occurring in food systems, Arce and Marsden (1993) urge the inclusion of what they call 'action at a distance': activity that begins outside the commodity system and that does not exclusively involve commodity producers. What follows is a description of the numerous actors who typically lie outside commodity production, but who engage in both material and cultural production activities. These actors are responsible for attempting to align production and consumption. As the fieldwork in subsequent chapters shows, they do not operate alone but enter strategic alliances or networks of exchange to achieve their respective ends.

ACTORS IN THE MIDDLE: RESTRUCTURING MATERIAL PRODUCTION

It is simple enough to charge that 'retail capital is increasingly mediating the producer-consumer relation' (Lowe & Wrigley 1996). While the claim is straightforward, the mediating mechanisms are anything but clear, with two alternative mechanisms discernible in relation to the producer-distributor interface. The more direct mechanism concerns mobilising the unique properties of retail capital. This alleged uniqueness relates to a number of factors: the nature and extent of retail capital concentration, limited oversight of retail capital by regulatory agencies, and retailer own-brands replacing manufacturer branded products. The indirect mechanism concerns the way in which retailers mobilise the capital they control to exert leverage over production relations, labour markets and supply chains: the latter referring to the distribution channels between primary and secondary producers, wholesale markets and retail traders. By taking charge in these three arenas, retailers drive the restructuring that is taking place between producers and distributors. What follows are some of the arguments being advanced in relation to retail capital and the restructuring of the producer-distributor interface.

CONCENTRATION AND OPERATION OF RETAIL CAPITAL

Consensus is emerging that the concentration of retail capital by a few firms confers purchasing power, and thus market power, on those firms (Burch & Goss 1999). This fact, coupled with its relatively rapid turnover, makes retail capital fundamental 'to the greatly accelerated circulation time of capital' (Lowe & Wrigley 1996, p. 9). The operation of retail capital begs wider questions about the adequacy of the Marxian political economy model, widely adopted to explain the circulation and conversion of different forms of capital. The most common explanation assumes that capital has different forms as it crosses three circuits: capital production, realisation and reproduction. In the first circuit, money capital is converted into productive capital, which purchases both labour power and the means of production. In the second, commodities are produced and are in turn exchanged for money capital in order to begin the process of further commodity production, which constitutes the third circuit (Bottomore 1983). Money capital which results from the sale of commodities has a specific function: to facilitate the circuit of realisation and, according to Bottomore (1983, p. 332), this is the moment of 'merchant capital', also referred to as retail and wholesale capital.

Retailers have long been confined to value-adding through moving goods between points: factory to wholesale operation and from wholesale point to retail operation. They have accumulated capital through 'engaging in repeated acts of exchange' (Ducatel & Blomley 1990, p. 216). Clearly this is changing. Large retailers are no longer content with reinvesting in their own share of activities and are either directly investing in productive activity and creating their own products or are specifying the productive investments of other firms (Burch & Goss 1999). Retail capitalists vie with productive capital for as much of the surplus value generated through exchange as possible and as a result intensify competition among producers and between producers and retailers.

Moreover, retail capital does not appear to attract government regulation to the extent of productive capital. This has been explained, in the context of Britain, by government reluctance to impede the relatively large capital flows and associated employment growth that has accompanied retail activity over the last several decades (Marsden & Wrigley 1996). In Australia, a recent parliamentary inquiry into retailing acknowledged an unprecedented degree in international terms of retail capital concentration, but it stopped short of recommending limits. The Committee defended its decision by arguing that this would be to the detriment of the hundreds of thousands of shareholders, including family and small shareholders, who had a stake in two of the major retail companies (Joint Select Committee on the Retailing Sector 1999, p. ix).

Whether retail/wholesale capital is a subform of productive capital with its own logic is the focus of debate. Ducatel and Blomley argue that retail capital operates according to a principle of 'separation-in-unity': it does not exist independently of productive capital but performs distinctively within the political economy and has 'a unique (and sometimes contradictory) logic within that ... larger system' (Blomley 1996, p. 238). Others argue that a sole emphasis on the economic functions of retail capital overlooks the way in which economic forces are culturally encoded (Lowe & Wrigley 1996).

Although the precise function of retail capital remains elusive, this form of capital is increasingly the subject of agrifood commodity analyses. For some commodity systems, the concentration of retail capital in a few firms influences the circulation of commodities as surely as the concentration of productive capital. What Chapters 5 and 6 show is that the oligopolistic conditions in the chicken meat supply chain — numerous product sellers (the ninety-odd chicken meat processors) and relatively few buyers (three major supermarket chains and a couple of major national chicken chains) — have a bearing on the fortunes of chicken. Under these conditions, retailers can dictate the product ranges that they are willing to sell and in this way influence the uses to which productive capital is put.

RETAIL RESTRUCTURING AND CAPITAL ACCUMULATION

I pointed out earlier that retail capitalists use their capital to invest in technologies and strategies that enable them to exert control over producers and, as we will see later, over consumers. Retail restructuring is the term that is given to a wide range of activities designed to position retailers at the forefront of commercial and social life. Lowe and Wrigley (1996, p. 7) identify a handful of developments which have influenced this restructuring, including organisational and technological transformations in retail distribution, reconfiguring labour practices within retailing and redesigning retail-supply chain interfaces.

Each development has been pertinent to the social life of the chicken and they are described below.

- Technological transformations in retail distribution

Over the last five decades, large food retailers have been the driving force behind changes to procuring and moving produce from farm gates to warehouses and onto retail shelves. Importantly, their role in improving the movement and storage of stock has allowed for even more rapid product innovation. Indeed, some argue that the advent of store and home freezers, followed by the cool chain (explained below), has been fundamental to the popularity of chicken (Symons 1982). In turn, it is highly likely that these technologies have enhanced the growth of supermarket power in the supply chain for this particular commodity as well as for other high value foods, including seafood, fruit and vegetables. One story will suffice to illustrate the mutually beneficial relationship between secondary producers, or processors, and retailers. Although the developments occurred in England and the United States, the technology and organisational transformations described were readily adopted in Australia.

According to Senker (1988), who undertook a detailed study of supermarkets' role in product innovation, the British supermarket chains Sainsburys and Marks & Spencer were instrumental in changing English consumers' views of chicken. By the mid-1950s, Sainsburys had become dissatisfied with its inability to handle large numbers of fresh chickens because birds that had been eviscerated remained saleable for only one to two days. Following a visit to the United States by representatives of Sainsburys and Poultry Packers, a processing firm with which it had a close relationship, the latter started to import both the freezing and evisceration technology to deliver a product range that did not have limited shelf life. Before long, Sainsburys was able to stock frozen chicken in their self-service stores, but this too was inadequate. Whole chickens were not considered by consumers to be a regular part of their diet, prompting further product innovation:

> There was a need for increased publicity to promote chicken in 1961 because although the technology of freezing chickens was well developed, the market lagged behind, even though the price of chicken had fallen and chicken had lost its 'luxury' label. At that time chicken was invariably bought whole, at the weekend only, and either roasted or boiled. A market research report considered that the market for chicken could be expanded by promoting chicken as a year-round staple food for instance by increasing sales of cut-up chicken during the week. Publicity had two themes (1) how to cut up chicken and (2) new recipes for chicken pieces (Senker 1988, p. 161).

Around the same time, Marks & Spencer was expending effort to find a suitable means of mass retailing chilled chicken. The result was what became known as 'the cool chain', since adopted across the industrialised world as a management technology used by supermarkets for procuring, transporting and storing fresh produce. In the case of chicken, it involved air chilling of eviscerated birds at the factory, transport in refrigerated trucks for travel to the retail stores and immediate transfer to chilled display cabinets. Market research showed that while the chilled birds were more expensive than frozen, housewives preferred them for their 'convenience'. They didn't have to thaw them, they tasted better, and the 'flood of water released by frozen birds during thawing made them seem a poor buy in comparison with chilled birds' (Senker 1988, p. 167). As a result of this supply-chain management innovation, Marks & Spencer became the recipient of the British Poultry Breeders & Hatcheries Association Marketing Award in 1979 for services to their industry.

This particular history of chicken meat processing and distribution highlights how supermarkets, rather than food processors, lay behind chicken's growing popularity in the 1960s. In Senker's words, 'innovations depended on large, capital-intensive retailers who had in-house technological capability to define the problems associated with the mass retailing of chicken, and to find appropriate solutions' (Senker 1988, p. 177). Since then, supermarkets have repeatedly introduced or demanded production system innovations (Burch & Pritchard 1996; Parsons 1996). Supermarkets, rather than consumers or producers, have been pre-eminent introducers of new food products and food practices into households.

- Reconfiguration of labour practices

Building on the work of the regulation school described in Chapter 2, David Harvey announced in 1989 that a new system of accumulation, known as flexible accumulation, had arrived. He described this system as directly confronting the rigidities of Fordism (Harvey 1989, p. 147). A year earlier in work rarely cited outside of retail geography, Murray (1989) had pointed out that it was retailers, not manufacturers, who were leading the industrial restructuring process that some would subsequently call flexible production and flexible specialisation. He argued that regulation theory took insufficient notice of the rapidly growing service industries and of 'circulation activities' because it was overly enthralled by product manufacturing. Murray justifies his claims by reference to the changing labour market practices adopted by retailers and their contribution to the reorganisation of 'consumption work'.

In contrast to what was happening in manufacturing, Murray identifies the retail and food service sectors as among the first to dispense with a full-time male workforce in favour of a part-time and casual workforce of women and young people. The adoption of a flexible labour market dramatically lowered wages bills and contributed to retailer profitability. Furthermore, he believed retailers were leading the way in simultaneously introducing employee multiskilling and deskilling. Murray's description of the 'revolution' in retailing applies without reservation to the Australian supermarket and fast food sectors over the last twenty years (see also Lyons 1996; Reeders 1988; Ryan & Burgess 1995).

In 1995 the Employment Studies Centre at the University of Newcastle (New South Wales) published several studies on the impact of labour market deregulation in female-dominated industries. In a study of supermarkets, the researchers found that the major changes to retailing practices revolved around changes in retail labour routines, including round-the-clock shifts, employment over unsociable working hours, an increase in part-time and/or casual arrangements and a requirement of skills related to information technology (Ryan & Burgess 1995, p. 33). The research highlighted the extent to which labour tenure was perceived by many in the supermarket workforce to be precarious, despite improved opportunities for employment on a permanent part-time basis as opposed to a casual shift basis. Workers voiced anxiety about their economic situation and their ability to cope with new job demands without adequate training. Furthermore, the researchers observed that the 'goodwill' of individual store managers was paramount to allow female employees to fulfil their family commitments (Ryan & Burgess 1995, p. 50).

In passing, the researchers noted that the feminisation and deregulation of supermarket labour markets and of trading hours occurred at the same time that supermarkets introduced new areas of employment: in-store bakeries, fruit and vegetable sections and value-adding to meats. The relationship between the trends of women's employment in the food services sector and household reliance on industrially prepared foods is not elaborated but is noteworthy, nevertheless, because it is an issue that pervades contemporary food-related research.

Mark Harvey (1998) makes more explicit links between labour markets and the market availability of products in an exploration of the practices of large British supermarkets. He found that the large supermarket chains coupled flexible labour markets with product differentiated retail systems. Such systems are the result of a single retail chain segmenting into different types of stores which contain distinctive product ranges, determined by consumer incomes and lifestyles: budget, own-brand, super-brand and niche products. A chicken and egg relationship exists because just as product differentiation is about responding to different consumer needs, supermarkets adopt product differentiation as a strategy to communicate to, and ultimately influence, consumer needs. Harvey argues that as retailers refashion working lives through their introduction of new labour processes, new employment contracts and new working patterns, they in turn create new markets for product ranges. Accordingly, supermarkets are capable of mediating production and consumption because of their ability to structure both 'the nature of exchanges and the social participants to the exchange' (Harvey 1998, p. 7). None of this would be possible, Harvey suggests, if supermarkets had not been able to respond to, and reinforce, patterns of product market and labour market segmentation simultaneously. Through their activities upstream in the supply chain and in their labour market activities, supermarkets have commodified food production, 'with profound consequences for the mode of consumption within the household' (Harvey 1998, p. 7). If Harvey is correct, it is possible to suggest that supermarket chains are usurping the role that agrifood producers have played in shaping consumption norms since the 1950s.[3] Supermarket chains appear, from his reading of the situation, to be co-ordinating key elements of the British food system.

- Reconfiguration of retail-supply chain interfaces

Mark Harvey (1998) is not alone in arguing that a major source of retailer power over suppliers and producers has come about due to their lead role in reconfiguring supply chains (see Gardner & Sheppard 1989; Marsden & Wrigley 1996). Australasian agrifood scholars have been busy in this area too (see Campbell & Coombes 1999; Parsons 1996). Burch and Goss (1999) recently nominated three changes that have resulted in supermarket dominance over the supply chains of high value foods: the global sourcing of product which forces competition upon local suppliers, the enormous purchasing power of the supermarkets which requires large volume suppliers and makes smaller firms redundant, and the movement into own-brand products that compete with products branded by manufacturers.

The combined effect of these factors has shifted the balance of power in favour of the retailers, as summed up in *The Economist*:

> The distribution chain used to be controlled by manufacturers and wholesalers. The retailer's role was to buy goods from the range offered by the wholesaler or other intermediaries, and sell them onto the consumer ... it was manufacturers who decided what goods were available, and in most countries at what price they could be sold to the public. That distribution system is now being turned upside down. The traditional supply chain, driven by manufacturer 'push', is becoming a demand chain driven by consumer [retailer] 'pull' (cited in Burch & Goss 1999, p. 336).

Three conditions underpin controlling the retailer-supplier chain: the designation of 'preferred suppliers', improved stock handling technologies, and retailer knowledge of the manufacturing process. For example, retailer entry into production through turning fresh into frozen chicken, followed by turning frozen back into fresh chicken via the cool chain, taught retailers what they could expect from suppliers. Their subsequent familiarity with production techniques, coupled with having alternative product sources, has meant that retailers can demand more of their suppliers. As a result, retailer specifications alter the mix of plant, labour and technology in which processors must invest. Furthermore, with the demise of small retailers, the suppliers have few, if any, alternatives to the large chains.

The development of relationships with preferred suppliers in Britain and Australia has been made possible by new management techniques and inter-firm contracting arrangements. One prominent form of relationship is termed 'relational contracting', which has developed in tandem with the growth in concentration of retail ownership (Foord et al. 1996). Relational contracting refers to contracts that are based on interactive, flexible and stable supply networks. While the last two features may seem contradictory they designate different temporal dimensions: the day-to-day orders may vary and are thought of as flexible but the contracts are ideally in place for a number of years, which is where stability enters. This form of contracting rarely involves formal, written contracts, but regular personal and telephone exchanges are used to negotiate price, recipe, quality standards and ingredients suppliers. It is the time demanded between production and delivery that induces inflexibility, possibly more so for suppliers with Just In Time delivery systems. Other contradictory tendencies are evident in relational contracting. It 'facilitates flexibility in meeting changes in demand and a reduction in the risk faced by the dominant partner in the relationship ... [but it creates] some inflexibilities — for example, the risk of excluding firms with "good" products' (Foord et al. 1996, p. 88). Thus, despite the 1980s being characterised by greater overall investments in flexible systems, this sort of contract is both intensive and centralised, as becomes evident from the research presented in Chapters 5 and 6.

ACTORS IN THE MIDDLE: RESTRUCTURING CULTURAL PRODUCTION

In her account of the importance of supermarkets to champion product innovation, Senker points out that even the best known food processors were unsuccessful at introducing new products because they did not know how 'to change conservative eating habits' (Senker 1988, p. 178). Supermarket capacity to encourage and respond to consumer demands through technology and market intelligence resonates throughout her study, as it does in Mark Harvey's work. Both researchers link the up-stream activities of retailers *vis a vis* producers with the down-stream cultural activities of consumers. A range of literature that illustrates the actions of retailers in concert with others to shape *the mouth of the community*, or the commodity context, will be drawn on in the present section, so as to appreciate the taste-making process at work.

What is noteworthy is the extent to which cultural production activities have involved actors from quite different sectors working together both strategically and opportunistically over many years. The systematic effort to entice people to shop differently, to patronise large stores rather than family-run small shops, and to turn more often to commodified foods has been going on for eighty years or more. The part played by advertising in encouraging these retailing practices has been critical to the commodification of the food system, as has been the use of psychographic research to assist advertisers to represent the commodity context in certain ways. Constant activity surrounds both the construction of commodity contexts and the status of commodities, and in the field of food one of the most potent symbols of status is a relationship to individual health. The role played by food in the generation of disease has been particularly significant to the esteem in which chicken is held. For this reason it is necessary to briefly describe the cultural process of assigning nutrient values to foods, or nutritionalisation.

At one level, retailing practices, psychographic research, advertising and nutritionalisation are independent practices. When each is put into the service of the other, however, they create a commodity context that has significant potential to influence consumption activities. The material that follows challenges those who think that consumer authority runs deep within the food system. It sheds a different light on food knowledge sources and shopping, and it reflects on the multiple processes that play a part in making food *good to think*.

RETAILING PRACTICES

Until the 1920s most food shopping in Britain and Australia was done from home with the help of the expert grocer who sent the delivery boy round for the weekly orders. According to Australian historian, Beverley Kingston, because housewives stayed at home 'there was little likelihood of impulse buying, of making comparisons, or encouragement to vary shopping habits' (Kingston 1994, p. 45). As a result, food retailing required advertising to communicate availability, and advertising became more important with the advent of branded products and choice between products in the interwar years. Advertising both led to the irrelevance of the skilled, knowledgeable and authoritative grocer and shifted the locus of decisions to a particular sort of woman: 'Mrs Consumer' (Friedmann 1990; Goodman & Redclift 1991; Humphery 1998; Symons 1982).

Eager to replace the grocer were self-service food stores, called supermarkets, which were built upon the concept of American department and variety stores. Australia had two highly successful variety store chains — Coles and Woolworths — which had begun trading in 1914 and 1924 respectively. By the 1950s their owners and senior employees were keen to introduce one-stop shopping for groceries as well as fruit and vegetables, frozen foods and delicatessen items. Both Woolies, as it is affectionately known, and Coles opened the doors of their first supermarkets in 1960. Making excellent use of historical company records, Humphery portrays what was obviously an exciting time for Australian retailers and shoppers. He reports for instance how one Woolworths' executive, returning from yet another trip abroad, extolled the virtues of adapting the 'American supermarket principle' to Australian conditions, thereby reducing 'the gap of progress estimated at twenty-five years between the two great "A's" — Australia and America' (Humphery 1998, p. 105). The American inspired 'revolution', as it was called by Sir Edgar Coles, continued unchecked with up to two stores a week being opened through the 1960s. The Coles chain was especially proud of its in-store cafeterias and showed little modesty when heralding its achievements. In describing the 1961 opening in Sydney of 'the largest and most modern Variety Store in the world', the publicity spiel boasted that:

> [e]ach of the many new departments had its special attractions, but perhaps the most outstanding is the new cafeteria. It ... provides facilities for the hygienic preparation and service of a great variety of foods under ideal conditions on a scale not previously seen in Sydney (Colesanco December 1961, p. 1).

These cafeterias became truly social places, for as one commentator pointed out:

> [e]veryone shopped at Coles at some time or another. Everyone met friends and family in one of the Coles cafeterias where you could get a plate of lamb's fry and bacon, mashed potatoes and peas for less than it would cost you to prepare it yourself. A bowl of apple crumble and a cup of tea rounded off a meal that was good value for money in anyone's language. In a sense, Coles was part of the way of life of working class Australians (Gawenda 1996, p. 29).

The research undertaken by Humphery (1998) highlights the continual recycling of ideas by supermarket chains in their quest to shape the retailing experience around their own needs for profit maximisation. He notes, for example, how convenience has been a mantra repeated by retailers for many decades and, more specifically, how from the outset supermarkets 'opted for the new attractiveness of the 'three C's': convenience; cleanliness; and consumer choice'. As a result, large food retailers have been critical to several discourses that permeate how we think about food systems more generally, including what is meant by convenience and how choice is judged. And they have had to deal with some major contradictions when getting consumers to behave in ways which accord with the supermarket view of how labour costs should be distributed. Arguably the most difficult period was obtaining acceptance from shoppers that they should handle and bag their own goods. In forsaking service for self-service the chains had to promote 'the image' of the new form of service which involved selling 'the free market system itself', as *Rydges Business Magazine* observed in 1964 (cited in Humphery 1998, p. 105). This they did with some enthusiasm and for this reason Humphery argues that supermarkets have been far more than food retailers. For more than half a century they have been viewed as significant social institutions for the way they represent the contours of national progress.

TRADE IN REPRESENTATIONS THROUGH ADVERTISING AND PSYCHOGRAPHICS

The role of advertising in the replacement of specialist food providores by supermarkets has already been noted but Kingston makes the point that supermarket supremacy was consolidated by the advent of television, 'though it was colour television that really bought shopping home' (Kingston 1994, p. 94). And while the process of advertising has remained unaltered throughout the 20th century, the process by which advertisers forge relationships with consumers has changed. Falk (1994) suggests that the now dominant form of advertising, involving audio-visual imagery, encourages much more emphasis upon the experiential aspect of consumption, rather than facts about the product. Image-based advertising lets producers and advertisers represent almost any promise (Robins 1994). In a commodity context characterised by an absence of falsifiable promises, advertising, it seems, eases the burden of accountability for both producers and consumers. But this form of communication comes at a substantial cost and could be argued to be an expensive form of value-adding.

Figures show that foodstuffs command a higher share of advertising expenditure than any other commodity in Australia, with global corporations the biggest food advertisers. In 1993, of the total expenditure on advertising spent by the top 100 advertisers in Australia (around $1.5 billion), close to $500 million was spent on food and beverage (including alcohol) advertising. Of this, McDonald's spent $45 million, Kellogg's $36 million, PepsiCo $30 million, Goodman Fielder $18 million (Australia's largest food company), the Australian Meat and Livestock Corporation (AMLC) $12 million, and the Australian Dairy Corporation $11.5 million (Sindall et al. 1994).

The sums spent by food producers are small compared to what retailers spend. Table 3.2 illustrates just how extensive retail advertising is. The table shows the biggest advertisers by various media for 1997, with Coles Myer appearing in all major media and Woolworths being a major advertiser in three of the four media analysed.

Table 3.2
Top ten advertisers by medium in 1997

	Metropolitan TV	Metropolitan newspapers	Magazines	Metropolitan radio
1	Unilever	Coles Myer	Nestle	Telstra
2	Telstra	Woolworths	Unilever	Publishing & Broadcasting
3	Coles Myer	Telstra	Toyota	Seven Network
4	Nestle	Optus	Telstra	Village Roadshow
5	McDonald's	Commonwealth Bank	Proctor & Gamble	McDonald's
6	Astre Automotive	Astre Austomotive	Woolworths	News Corporation
7	PepsiCo	Incape Motors	Astre Automotive	Woolworths
8	Mars Inc	Mitsubishi Motors	Coles Myer	Ten Network
9	General Motors-Holden	Qantas	Mitsubishi Motors	Optus
10	Cadbury Schweppes	Village Roadshow	Qantas	Toyota

SOURCE Shoebridge 1998, p. 68.

Despite advertising being a relatively non-threatening form of communication, the business of how to advertise food is a highly charged activity. Because food is burdened with a symbolic load, those who are involved in food communications are forced to adopt a dual emphasis upon the internal (personal) and external (social) meanings of food. For this reason Mintz argues that those who codify food are worthy of critical analysis (Mintz 1994, pp. 114–15).

The largest corporations base their communications about themselves, their products and services on extensive knowledge of consumer fears, prejudices, desires and behaviours (Tansey & Worsley 1995, chs. 7 & 8). An entire field, consumer sciences, has evolved to service the needs of corporations keen to martial understandings of human behaviour and emotions. Research techniques such as focus groups and psychographic mapping have been especially important bases for marketing and advertising, which bring products to the consciousness of consumers. In a detailed study of the way in which the Tavistock Institute of Human Relations in England worked, Miller and Rose (1997) provide insights into the economic value of psychographic research. While this particular research centre was established to apply theories of human relations to the advancement of social ends, its major achievement was to improve the use of behavioural and social psychology tools to aid commodification. Its 'psy expertise' was sought by companies keen to develop marketing and advertising strategies to sell their goods. Among the Institute's credits in the food area were: increased sales of ice cream in winter; helping consumers to overcome the guilt that accompanied eating chocolate; and increased meat consumption through teaching the uses of mustard.

Miller and Rose argue that the Tavistock Institute was able to mobilise consumers 'by forming connections between human passions, hopes and anxieties, and very specific features of goods enmeshed in particular consumption practices' (Miller & Rose 1997, p. 2). They add:

> [the] work of the [Tavistock Institute] is thus characteristic of a wider set of processes that were involved in shaping the 'commercial domain' in the mid-20th century and the beliefs and forms of conduct that made it up ... This was not a matter of the unscrupulous manipulation of passive consumers: technologies of consumption depended upon fabricating delicate affiliations between the active choices of potential consumers and the qualities, pleasures, satisfactions represented in the product, organised in part through the practices of advertising and marketing, and always undertaken in the light of particular beliefs about the nature of human subjectivity (Miller & Rose 1997, p. 34).

While producers and retailers have relied for many years on the sort of 'psy expertise' which they pay others for, supermarkets have been gathering their own knowledge of consumers through electronic point-of-sale systems and loyalty programs. Indeed, their consumer intelligence has become so great as to prompt the proposition that retailers are revising 'the rights to consume ... while simultaneously attempting to define consumption interests around their own particular notions' (Lowe & Wrigley 1996, p. 11).

If, as some argue, advertising has replaced education in showing consumers how to act appropriately (Warde 1994), the amounts spent on advertising by supermarkets and fast food chains to assist consumers to participate in consumption activities becomes significant. As part of their study of microwaves, Cockburn and Ormrod (1993) confirm how the mass media has been displacing mass education in helping to construct hierarchies of values. They argue that the teaching of home economics in schools has been subordinated by 'technoscience' and commerce, both of which are massively promoted in the media. The importance of the mass media in relation to judgements about food becomes clearer when the communication of nutrition science is considered.

NUTRITIONALISATION

Food consumption in the 20th century appears from sociological and anthropological accounts to have been influenced by two major 'foodways': traditional culinary culture as transmitted through kinship networks, and professional/quasi-state sources such as home economics, dietetics and nutritional medicine. However, it appears that the second pathway is assuming increasing significance, promoted as it is through popular culture, including advertising. At the same time, direct personal knowledge of foods' origins and composition are being diminished through reliance on an industrial and global food supply. Not only has food production become progressively delocalised[4] and technology driven, but food consumption is increasingly disembedded from traditions and customs.[5] At a time when consumers are distanced both geographically and metaphorically from the production process, attention given to the health attributes of foods is now greater than ever. The result of advances in scientific understanding of the diet-disease relationship, widespread dissemination of knowledge in the mass media and a focus on the body in the construction of identity all lay claim to the spotlight.

Thus, consumer concern about food values is magnified at the very time that the food supply is at its most remote from daily experience. This paradox heightens perceptions of anxiety, risk and vulnerability, manifesting in food scares and panics. Indeed, the focus groups conducted as part of this research confirm that while consumers take food availability for granted, they are not relaxed about food's meanings or worth. This reverses the longstanding food ontology for older, industrialised societies, which was concerned with obtaining enough food to survive.

Where experiences of risk from food related disease are heightened through the media and where product diversification meets omnivore's paradox, as discussed in the previous chapter, the food industry has embarked upon re-embedding trust in the food supply. To this end, the nutritional value of food is mobilised. Health claims attached to foods have subsequently become an important ingredient in the fight for competitive advantage (Nestle 1993) and numerous analysts have observed how nutritionists have been co-opted by food producers (Belasco 1989; Levenstein 1993; Nestle 1993). What these North American and other Australian studies (Crotty 1995; Santich 1995b) do is to emphasise the importance of science to support partisan appeals that food products and particular dietary practices are *good to think*.[6]

The importance of the second foodway is supported by market analyses. In one study by Australia's largest scientific organisation, the CSIRO found that the most reliable sources of nutrition were, in order, the National Heart Foundation, the Anti-Cancer Foundations and dietitians, and doctors. The least reliable were food labels, mass media, food manufacturers and food advertisements (CSIRO 1994, p. 15). Similar findings were produced in research commissioned by the Australian New Zealand Food Authority.

Such evidence may explain why retailers are beginning to follow producers in legitimising their activities through associations with nutrition science.[7] In mid-1999, Coles supermarkets and the Dietitians Association of Australia co-sponsored the 7-A-Day Campaign: a push to encourage the daily consumption of seven serves of fruit and vegetable.[8] The Federal Minister for Health and Aged Care lent his authority to the launch via a press release. In that document, the Coles Managing Director was quoted as saying '… Coles believes it has a responsibility to contribute to improved health in the community through better information and advice in partnership with nutrition experts' (Anonymous 1999). This is a recent Australian example indicating the increasing subjection of food systems to the process of nutritionalisation or, what has been termed elsewhere, 'nutrification' (Belasco 1989). It is a particular form of value-adding that entails shaping the consumption discourses around consumer health concerns.

EXPLICATING DISTRIBUTION AND EXCHANGE ACTIVITIES

The foregoing rather fragmented knowledge base about material and cultural mediation of production and consumption leads me to propose the following processes and agents as important to the spheres of distribution and exchange:

- *supermarket retailing practices and organisation*: requires the understanding of retail capital concentration among the dominant retailers, including the way they are organised to influence the exchange process between suppliers, themselves and consumers

- *food services sector organisation and practices*: acknowledges the special and growing function of ready-to-eat food supplied by caterers and institutional food outlets such as workplace and school canteens, restaurants and fast food outlets

- *retailer-led product development*: acknowledges increasing in-house product development and preparation, and the pressure on suppliers to deliver particular products

- *marketing and distribution networks*: as Friedland has described, but with emphasis on the increasing integration between producers and distributors through supplier contracts and Just In Time distributive processes. Consideration can be given to the use of psychographic and similar consumer research to position goods and services in the marketplace.

- *labour as a factor of distribution*: includes those paid to move food from the farm gate to the processor, market/warehouse and retailer as well as those working in retailing and food service
- *food knowledge and discourse production:* acknowledges the actors and activities which comprise diet-making activities as described above, including government employees and academics who propose dietary guidelines, food journalists and gastronomes, corporate advertising and corporate lobbyists, nutrition science research, food and wine festivals and cookbooks. It also extends to food and animal welfare activism.
- *regulatory politics*: involves the government policy process in relation to issues such as land-use planning and retail sites, laws regulating retail opening hours and labour markets, advertising codes, health claims legislation, consumer protection and retailer competition rules.

ORGANISATION OF THE FIELDWORK

The amended CSA framework is summarised in Table 3.3. The headings are used to organise the material that follows in Chapters 4 to 7.

Table 3.3
Amended commodity systems analysis framework

PRODUCTION PROCESSES

- Primary and secondary production practices
- Grower organisation and organisations
- Labour as a factor of production
- Science production and application
- Product design
- Regulatory politics

DISTRIBUTION PROCESSES

- Supermarket retailing practices and organisation
- Food services sector organisation and practices
- Retailer-led product development
- Marketing and distribution networks
- Labour as a factor of distribution
- Food knowledge and discourse production
- Regulatory politics

CONSUMPTION PROCESSES

- Tertiary production practices
- Means of access
- Manner of delivery
- The eating environment
- The eating experience

CONCLUSION

This chapter outlines a new model for analysing the social life of a single commodity, with a view to illuminating the balance of power between producers and consumers. The model acknowledges the input and interests of a range of actors, including those who lie outside the agricultural and household sectors and emphasises value-adding processes beyond the sphere of production and consumption. Retailers figure prominently as playing a part in value-adding.

4

CONSUMING CHICKEN: BUYING TIME, NUTRITION AND FAMILY HARMONY

> As an antidote for colds and flu, depression, bad report cards, upset stomachs, cramps, political unrest, allergies, bronchitis, arthritis, and hangnails, a bowl of steaming hot chicken soup is cherished in most every part of the world ... If in fact 'Jewish penicillin' doesn't solve life's most stubborn problems, it at least relieves the symptoms.
>
> (Hazen 1994, p. 1)

> I could cook chicken a different way each night for a long time without becoming bored, and I am never unmoved by the sight of a roasted chicken. In the restaurant, however, my chicken dishes are never best sellers. It seems the public equates chicken these days with the ordinary, the everyday and the cutprice.[1]
>
> (Alexander 1996, p. 207)

Now that the themes, issues and major concepts have been laid out, this chapter begins to unfold the fieldwork. I start with consumption sphere activities because, as Chapter 2 reveals, one of the preoccupations of the sociology of consumption concerns the power of consumers to shape what is produced. In some accounts, consumers achieve their influence over the market through *not* purchasing goods and services (Douglas 1997; Miller 1995). In others, consumer sovereignty results from consumers constructing, or adding, values not intended by the product creator (Falk 1994). As a consequence, consumers are said to enhance a commodity's economic value through attaching values to goods and services (Bauman 1988; Warde 1992). Moreover, women are portrayed as being particularly empowered in consumption given that here *she* does the work and furthermore excels compared to men. It is *she* who challenges the state and capitalism to improve their moral and market economies (Douglas 1997; Miller 1995).

I was curious to know to what extent consumers believe these claims when even the more enthusiastic proponents of the notion of powerful consumers acknowledge the contingent nature of that power. Falk (1994), for example, distinguishes between passive and active consumers while Bauman (1988) talks of the seduced and repressed consumer. Both authors argue that consumers differ in their capacity to influence the workings of the market. Furthermore, what one learns from the gendered micro-sociologies of feeding the family is that food consumption is a matter of getting by, of making do, and of resignation that the market does not necessarily deliver what is wanted (Charles & Kerr 1986; DeVault 1991; Murcott 1982; 1986).

This chapter describes what a group of Australian consumers actually say about their power in relation to the food system and, more specifically, chicken meat consumption activities. The primary consumption data was gained from five focus groups, in total containing thirty-three men and women in metropolitan Melbourne, conducted during 1996–97. Three of the groups consisted of parents who met as playgroup participants, another group consisted of workers from a common employer and the fifth consisted of a mix of work associates and their friends, all of whom were employed. I call these two groups 'the workgroups' as distinct from 'the playgroups'. As a result of a personal data sheet completed by participants, each of the groups was ascribed a particular socio-economic status (SES).[2] Two of the playgroups were designated as lower socio-economic status, the two workgroups as middle socio-economic status and the third playgroup was best classified as lower-middle class, or sitting somewhere in between the other groups. The insights gained from the focus groups are remarkably consistent with market research conducted for commercial purposes and referred to later in the chapter.

Before proceeding to outline the material organised by way of the amended commodity analysis framework described in Chapter 3, I provide an overview of chicken meat consumption in Australia followed by a section entitled 'Feeding the family'. That section summarises the focus group participants' views of food provisioning in the context of family and working lives.[3]

CONTEMPORARY CONSUMPTION OF CHICKEN

Australians have long been a nation of meat eaters and, until the 1970s, the preferred choice of housewives reflected the farmers' offerings of beef, lamb and mutton. Indeed, there has been a clear symmetry between the mass production and consumption of meat for fifty years, with price being the major determinant behind all but the richest households' decisions to switch between meat products (Larkin 1991; Turner 1977).

Prior to World War 2, Australia's per capita meat consumption stood at 118.5 kilograms per annum, of which beef comprised approximately two-thirds, with mutton and lamb constituting the remainder. Meat consumption peaked in the mid-1970s (124 kilograms per capita), yet by 1978 beef had dropped to just over one half of the total, with poultry and sheep meat consumption comprising the remainder in almost equal parts. Poultry has continued to outpace lamb and, on average, currently comprises thirty per cent of every Australian's meat consumption.

According to the Australian Chicken Growers Council, Australia is the fourth highest chicken consuming nation in the world. On the basis of both rising chicken consumption over the last thirty years and its supply by a small number of chicken growers and processors, the Australian poultry industry claims to be the country's most successful agrifood industry (Australian Chicken Meat Federation n.d.; Blackett 1970).

Table 4.1 compares per capita consumption of different meats since the 1950s in Australia. The data confirm poultry's upward consumption trend and support a prediction that Australia's per capita chicken consumption will overtake beef by 2013, a trend which is shared with other industrialised nations (Larkin & Heilbron 1997).

Table 4.1
Australian meat consumption per person (in kilograms)

Year	Beef and veal	Mutton and lamb	Pig	Poultry[4]
1958–59	56.2	36.4	4.6	4.4
1968–69	40.0	39.3	6.7	8.3
1978–79	64.8	18.0	13.3	17.1
1988–89	40.0	22.2	17.5	24.1
1998–99	36.4	16.3	19.0	30.8

SOURCE Australian Bureau of Statistics 2001.

POWER EXPERIENCED WHEN FEEDING THE FAMILY

Consistent with previous studies from Britain, the United States, Canada and Australia, the focus groups for this research contained women who, regardless of whether they were employed for a wage, took responsibility for meal planning, shopping decisions, meal preparation and serving. There were a handful of exceptions. In one household the woman hated cooking and her husband did more than half. When they shopped, each made their own purchasing decisions. In three households the men did an equal amount of food shopping, but on the basis of lists prepared by their wives. In a further household, the man, a former butcher, made the meat and fish purchases. In one of two instances, where the young couple was childless, the man and woman shopped and cooked together.

These exceptions amount to only six of the thirty-three households. In this regard the focus groups reflect the general situation that prevails in regard to the gendered nature of household food work reported by Bittman (1992). His Time Use survey data demonstrates that while men are doing more cooking than in the 1970s, and women are doing significantly less, women still do far more than men. Furthermore, women continue to do most of the shopping. Bittman notes that while the overall time spent on household duties has barely changed in the last century, marked shifts are apparent in the way that time is allocated and the way tasks are broken down. For example, less time is spent cooking and cleaning but more time is spent shopping and storing food. In other research that uses Household Expenditure data, Bittman and Mather (1994) have also shown a significant shift in cooking from scratch to the use of pre-prepared meals. What their plausible analysis does not do is to canvass how the social actors themselves view these activities and changes.

The statement 'I love food, but under the right circumstances' would best summarise the consumer sentiment encountered in my discussions. Discovering, however, what constitutes 'the right circumstances' proved elusive, thus confirming the findings of those studies that are more sceptical of consumer sovereignty and authority. Specifically, meal planning seemed universally disliked and regular food shopping, especially shopping with too little time and with young children, was considered 'a pain' by almost everyone.[5] Cooking was generally rated as okay, enjoyable and creative except when constrained by time and children (which was most of the time). For some, juggling husband's and children's preferences was the major constraint to enjoying cooking. For those who were expected to cook a meal every night after work, cooking was experienced as 'a chore', with weekend cooking offering more promise for those in the workforce and with older children.

Furthermore, the focus groups provided a very mixed assessment of women deriving power from, or exercising power over, food-related activities. Whether in paid full-time employment or not, women participants asserted that feeding the family was their domain in spite of enjoying few of the duties that comprise food provisioning. Like Counihan (1988), whose research was reported upon in Chapter 2, I did not see any great desire for this to change, however. This was particularly manifest in the playgroups where the women were reluctant to give over shopping activity to their male partners. Indeed, they supported Campbell's observation that '[t]he ability to present males as effectively 'incompetent' at shopping enables women to argue that men should not be allowed to engage in the activity and hence to volunteer (sometimes with a mock show of reluctance) to do their shopping for them' (Campbell 1997, p. 173). However, the groups did not lead me to believe that women behave like this because they are aware of the power they exercise over both markets and men. The women with less disposable income controlled shopping out of a sense of thrift and the professional women seemed concerned about the poorer quality foods that men purchased.

The Melbourne-based consumers highlighted another layer of complexity: deference towards children's demands. Children it seems are as potent a boundary setter for what food comes into the house as is deference to men. This was especially apparent when the households without husbands contained women who were still not consistently eating their preferred foods. Rather they were acquiescing to the sensitivities and desires of young people. This was also apparent in discussions of how teenagers' changing tastes needed to be accommodated in the quest for family harmony. The possibly unprecedented role of children's preferences over household dietary practices has been highlighted by Gofton, who argues that:

> [c]hildren, from having their subordination and dependence expressed in their subjection to the disciplines of meal times, and their appetites regulated into the taste for normal adult foods, taken in normal adult ways, have been allowed far greater autonomy — indeed, they form the focus for much household activity (Gofton 1990, p. 92).

McIntosh and Zey (1989) conclude their own assessment of women's power over food provisioning by arguing for a need to distinguish between power, authority, coercion, influence and control. They suggest that women do not possess power or control rather they 'possess potentially powerful resources' including labour force participation, status production, emotional or sexual manipulation and control over household technology.

Given the proposition that women derive less self-esteem from meal-related activities, than in the (unspecified) past, it could be expected that women would welcome opportunities to side-step domestic labour duties if they had the chance (Gofton 1990, p. 92). This is not supported by the focus group research. For most of the women participants, satisfying other's needs and desires appeared both to be an accepted and an important part of their lives. In her ethnographic study of household life in South Australia, Duruz, in my opinion, is better able to capture the dynamics of competing behaviours and emotions when she notes:

> [d]iscourses of comfort may build a bulwark against capitalism's high powered ideologies of consumerism and the accompanying forms of

> the patriarchal division of labour, and may open up a space for personal pleasure and fulfilment. The satisfaction of facilitating others' comfort may seem to offer some compensation for the ambivalence of positioning in household power relations, and in structures of power more generally (Duruz 1994, p. 107).

The use of food to provide comfort was evident in spite of most focus group participants giving consistent expression to the idea that being a food decision-maker is a thankless and stressful task. They indicated that participation by outsiders in making decisions is greatly appreciated especially in light of the degree of anxiety that food matters evoke. The range of food anxieties disclosed was extensive: fear of food poisoning, dietary related health problems, and how to evaluate food claims of goodness comprised the bulk of the responses. The first anxiety is not new and the second and third have been a feature of middle class life for several decades (Crotty 1995).

Anxiety over family well-being and healthfulness led all the focus group participants to desire more nutritional information and assistance with food selection, as long as such assistance did not come from food corporations. Ideas for meals were also appreciated: a point not lost on corporations who are aware that they must offer 'meal solutions' (ASI/AC Nielsen 1998; Steggles 1996). Generalised anxiety about the food supply and 'nutritional cacophony' is the context in which social scientists have noted a dramatic increase in the consumption of cookbooks, television food programs and dietary advice (Fischler 1993; Giddens 1992).

CHICKEN MEAT CONSUMPTION SPHERE ACTIVITIES

The remainder of the chapter provides a consumer perspective of chicken meat, within the more general context of family food provisioning and culinary culture. It attempts to unravel the significant range of sentiments toward chicken which are reflected in the opening quotations: its mystical qualities, nutritional value, evocative spirit, culinary adaptability, and value as cheap human fodder. Through both the primary and secondary data I have sought to understand why chicken has been assuming a more important role in household diets. I have also attempted to establish whether chicken meat consumption raises any concerns for consumers. Food concerns are important in terms of exploring consumer power over individual food decisions, over food production more generally, and for obtaining insights into which regimes of value currently underpin Australia's culinary culture.

TERTIARY PRODUCTION PRACTICES

Since the 1970s, chicken meat products have been prepared in four sites: processing plants, supermarkets, the food service industry (which includes fast food and take away outlets and restaurants), and household kitchens. By constructing the social life of chicken out of the experiences and understandings of the focus group participants, the impact of the dramatic shift from household to industrial production for this particular food becomes apparent.

Up until the 1960s it would have made sense to describe the home backyard production process, where family members would kill and 'dress' the chickens they had raised. Many focus group participants spoke spontaneously of their memories of this process, and of their feelings of seeing one of the 'family pets' sitting on the table in a roasted form. The following story as recounted by Marty, whilst more graphic than many, was told to me in various ways by a third of the participants and by countless friends and acquaintances over the course of the research:

> My parents had chooks when I was growing up, and occasionally we killed one to cook it. It was the best entertainment when Dad wrung its neck and then he'd tie string around its legs and hang it from the clothes line and then Mum would have the laundry tub full of really hot water and, once it had drained, in it would go and she would pull out the feathers, and hand up bum and pull out the innards and we saw it and smelt it and knew what was happening and it didn't worry me one bit, and then Mum would stuff it and cook it and we'd get stuck in. We accepted that was life, that part of eating the chook was death. Then I grew up. Blah, blah. Now, there's no way in the world that I'd roast a chook. Some years ago I used to roast a chook from time to time, and then I would always buy a chook already cooked from the cut price deli, 'cos they do a really good one. But in recent times I've heard too much about what goes into chooks when you buy it, so now there is really little chook in the diet now. So when we get Chinese take away we'll get chicken satays. But I wouldn't serve big bits of chook anymore.

Among other things, Marty is pointing out that for most people, chickens are now raised and killed by some anonymous other, the situation pertaining to cattle and sheep over the 20th century. What Marty's story does not illustrate is that unlike beef, sheep and pork products, the majority of chicken is purchased ready-to-eat, or in a heat-and-serve form. This is the sense in which the processing and retail industries talk about the *convenience* of chicken — if chicken does require home cooking, it will be an easy and brief cooking episode (Steggles 1996). For a food that has attracted the title of convenient, however, its preparation for consumption caused some family cooks a deal of discomfort.

While I did not ask the focus group participants to compare chicken's contemporary preparation with other meats, it seemed to cause more distress. Jenny, from one of the playgroups, and for whom chicken in the family diet is 'extremely important', told of the emotional upset caused through its handling. At one point she said, 'I hate chicken Maryland, too much fat in it', and later she revealed, 'I was in tears in our first month of marriage, I was there with surgical gloves, cos I used to have eczema — just to pluck the feathers off'. Similarly Tracy, whose diary revealed chicken meals on five out of nine days, commented: 'I bought chicken yesterday and it was disgusting — the Maryland, I took the skin off and it was full of fat. I stood there for nearly half an hour picking it off. It was terrible'.

The lack of knowledge in respondents concerning the origins of chicken meat and what is involved in its production is notable, considering the short space of time in which production has gone from the backyard to the processor and retailer. The following conversation held in one of the playgroups sums up the state of most focus group participants' knowledge of the primary production process. It also illustrates the dilemmas consumers experience when thinking about their place in that process:

KAREN: Well I've driven past a chicken place, and that's where they come from. Do they keep them like battery hens?

CHRIS: I assumed they're farmed just like cattle and brought to a central place to be processed.

STEVE: A chicken battery system where they're fattened up, cut and processed, 'cos everything's based on economics. I remember caponising chickens to fatten them up. People don't want to know about it.

KATE: No I don't.

SHIONA: I have no idea and no desire to.

INTERVIEWER: In case it puts you off chicken?

SHIONA: Yeah.

JEAN: You know all about the hormones that're meant to make them bigger and fatter. I think about it, I honestly do. I think about growing up on a farm where we raised chickens and chopped of their heads, and ducks and pigs. They were well looked after. Yes, I do get concerned about what we are getting fed and we don't know anything about it.

SHIONA: Yes, I like to think they're all on farms and have a nice life before …

CHRIS CUTS IN: With this BSE problem in the UK the frightening thing that came out is that it was processed sheep that the cattle were eating — it was very surprising where by-products went — it goes through the food chain.

STEVE: Then you start to think about the grains that have been fed to the chickens, and they've been fed with all the pesticides and the fancy chemicals and so forth, so you're getting them anyway, and its the same with all your meats. So you just ignore it, and don't think turning vegetarian will solve it either!

MEANS OF ACCESS

In his consumption sphere schema, Warde (1992) emphasises the process of exchange. Three of his four exchange relationships were present in the focus group discussions: market exchange, familial obligation, and to a lesser extent reciprocity.[6] This last form was most evident in the workgroups, when the talk turned to having friends around or having dinner parties.

The shift away from a mix of exchange relationships towards almost total reliance on the market has been particularly apparent with chicken. Chicken meat for this group of thirty-three Melburnians is now exclusively procured in the market. In the space of less than one generation, the exchange of chicken meat for money has replaced the self-provisioning from the backyard chook house, as well as the reciprocal exchange of chicken for some other home grown produce.[7] This finding is not a surprise when one considers the labour, time and skills required to be self-sufficient in meat, chicken meat included.

The more involved chicken commodity chain, incorporating new actors like retailers, was not viewed as a bad development in the present study. As indicated above, wishing to remain ignorant of the life and death process of the chicken was a common response. Other comments which illustrate this attitude to consuming chicken included general agreement in one of the workgroups that, 'I try not to think (about where chickens come from) otherwise I'd never eat it'. In one playgroup a mother stated, 'No, I don't know about chicken 'cos if I knew I probably won't eat it'. This view was shared in the second workgroup, 'I love chicken, and I've just realized I'm not prepared to read the true story 'cos I think I might have to give it up'. Another person in this group possibly encapsulated the sentiments across all five groups when she reflected that 'I know that I don't want to know in some respects because I can't afford to know. I just don't have the resources at the moment to deal with it all'. For this highly educated single mother, the resources she was referring to were financial. However, for others 'the resource' would be the effort of finding an alternative that provides the same 'meal solution' for feeding the family that chicken provides.

MANNER OF DELIVERY

Only two of the respondents mentioned purchasing their chicken meat from a specialist poulterer, and one supported this in her diary. Several said that they would use this sort of outlet more often if they were more conveniently sited. Overwhelmingly, the supermarket is the primary site of purchase of chicken meat. Table 4.2 shows how supermarkets have over this decade been displacing chicken shops and butchers as the major source of fresh chicken purchases (Steggles 1996).

Table 4.2
Share of fresh chicken customers by outlet 1992–96*

	June 1992 (n = 1328)	May 1993 (n = 1315)	July 1994 (n = 1425)	July 1995 (n = 1469)	Sept 1996 (n = 1426)
	%	%	%	%	%
Total supermarket	44	48	57	61	62
Woolworths/Safeway	14	17	22	24	26
Coles	17	16	16	18	16
Franklins	3	1	3	4	4
Bi-Lo	3	2	3	4	4
Other supermarket	7	12	13	11	12
Chicken shop/delicatessen	29	23	21	20	18
Butcher	20	23	17	14	15
Other	7	6	5	5	5
TOTAL	100	100	100	100	100

*National figures, all fresh chicken buyers (Steggles 1996).

For most, supermarket shopping requires access to a car. Except for one older participant who relied on her husband to drive her shopping, no one mentioned transport access as a factor when shopping.

A further key feature of supermarkets is the necessity for consumers to largely self-serve. As has already been described, the constraining factor referred to most frequently in regard to food shopping was the presence of young children, and much planning was undertaken to leave young ones at home with the partner. If this was not possible, one-stop shop convenience, being able to dash-in-and-out and find items quickly, was of utmost importance. This is where the use of shopping lists was stressed. They were not only used as budgeting and planning tools, but as delivery devices: know what you want, go to the right aisle, find, pay, and exit. For this reason 'strange supermarkets are a pain'. The dilemma voiced over shopping in this manner was the way it precluded shopping around, reading labels, and exercising substantive choice. Supermarkets are aware of these dynamics and it is no accident that the rotisserie chicken section is placed near the entrance of newer and refurbished supermarkets.[8] This allows for dash-in, dash-out meal delivery.

THE EATING ENVIRONMENT

Most chicken was consumed with others present and according to the diaries I collected, the majority of chicken meat meals were consumed in the respondents' own house. The exceptions were the two groups where the women (and men) were employed, for whom chicken sandwiches and rolls and chicken and chips were consumed for lunch.

These were mainly purchased near the workplace.

Given the 'cooking-after-work' factor and the general appreciation of being cooked for, the issue of having the food services sector provide the family meals was explored. The market research firm, BIS Shrapnel, reported that in 1994 Australians spent one in three of their food dollars on eating out and take away meals (BIS Shrapnel 1995). This figure amounted to an increase of fifty-eight per cent over the previous five years. The analysis showed that the fast food component accounted for just over half of this expenditure, with the remainder being spent in restaurants, cafes and hotels.

The focus group discussions and the returned diaries support a proposition that for those with less disposable income and small children, eating out — except at other family member's homes — does not happen on a regular basis. Nor does bringing home take-away meals. It was primarily those in the workforce who mentioned bringing home ready-to-eat pizza or rotisserie chicken as a substitute for a home cooked meal. This is hardly surprising because of the greater time constraints and higher incomes of these groups and as pointed out in the context of Britain, convenience foods require a second wage earner (Goodman & Redclift 1991, p. 11).[9]

Those in the playgroups discussed having a KFC meal, for example, as 'a treat' rather than providing a home meal replacement strategy. This was clearly illustrated in one of the focus groups, which was conducted as part of the playgroup's Christmas break-up. On this occasion the mothers had supplied KFC nuggets for the children 'as something special'.[10] A participant in another playgroup also mentioned that KFC was present at their Christmas break-up party. In these instances the fast food did not replace home cooking, but supplemented it.

When fast food does replace home cooking for lower SES groups the food is consumed in the fast food restaurant. Mackay has pointed out that the attraction of chains like McDonald's lies in the way they position themselves as 'community kitchens': a local restaurant where fast food combines with slow eating (Mackay 1992, p. 11). Such restaurants provided lower income participants with an opportunity to eat out with young children, and not have to pay for a babysitter. These establishments were considered more tolerant of children and more affordable than restaurants. One participant commented, 'No, I'd never take Natalie to a restaurant, she'd never be patient enough — to McDonald's or something, yeah', while another lamented that, 'a proper restaurant — not as frequent as it used to be but the fast food places like a pizza shop, or casual places, yeah we'll go there more frequently, about once a month'.

At the same time, participants from all groups reacted against take away food and fast food. In the groups with young children there was a concern for the health implications of fast food. In the lower middle-class playgroup home cooking versus take away meals was thoroughly evaluated. Steve said that they cooked in their family 'because of concerns with the quality of the food that's the fast food', to which a chorus added 'health' and 'expense'. Karen remarked that, 'Our kids love pizzas and for the reasons above [cost and health] we make our own', but Shiona cut in saying, 'I used to make my own occasionally, but there's a place in Ferntree Gully now that does them for $4 and I can't make pizzas for that'.

THE EXPERIENCE OF EATING CHICKEN

Two chicken consumption experiences were common to all groups. The first was concern about the healthfulness of eating chicken and the second was easy of choice because it was relatively uncontroversial in the family diet. Yet few of the participants unequivocally enjoyed chicken. One could surmise that without a family to feed, chicken consumption would fall dramatically. However, it continues to be a festive food reserved for important occasions (Root 1980) and one imbued with special connotations as the Hazen quote which opens the chapter highlights. Simultaneously it has become classed as an ordinary food as Alexander's observation at the beginning of the chapter indicates.

Given that it now comprises more than one third of Australians' meat consumption, chicken, somewhat surprisingly, seemed to evoke varying degrees of anxiety from all focus group participants. They were most concerned with hormones and additives to the birds' feed. They feared ingesting the additives via the chicken and becoming sick as a result. Three of the participants mentioned eating less chicken for this reason.

In addition, the great majority of respondents viewed chicken meat with ambivalence — by this I mean that two-thirds nominated both good and bad characteristics for their consumption of chicken. I outline below the diversity of reasons given for buying, preparing and eating chicken. It should be noted that most who nominated a characteristic they liked about chicken followed this with something that hampered that enjoyment or caused them concern.

- Personal liking for chicken

At least half a dozen women mentioned their personal 'love' of chicken and its taste. However, each qualified this particular statement with an observation that another family member — husband or child — didn't share her view. One of the women remarked, 'I love chicken so its pretty high up, but I'm also conscious about not missing out on my iron in red meat, and my husband loves red meat so I try and get a balance, chicken tonight, red meat tomorrow'. Another lamented,

'Chicken is the food I'd go for. When the kids are home they eat a lot of lamb [and] sausages'. While another in this group of mainly single mothers said, 'I'm sorry that we don't have so much chicken since the vegetarian descended. I no longer buy a whole chook and use the chicken bones to make chicken stock — which I think is a wonderful thing'.

In one of the playgroups, a Maltese woman explained why her household did not eat more chicken: 'Because my husband doesn't like chicken ... he likes fish; my house is where my husband wants more red meat'. Later, Tracy in this group said, 'I've gotta cook chicken or fish, if I make a vegetable dish he feels as if he hasn't eaten anything' and Rita rejoined, 'Same here'.

As with most meals, children influenced which chicken meals were prepared. Surrounded by their play-schoolers, mothers discussed their own favourite chicken meals. One said, 'Roast chicken [is our favourite chicken meal] and chicken cacciatore, but she [pointing to her young daughter] won't eat what we eat — she won't eat anything new, so I'll do whatever she wants 'cos otherwise she won't eat it'. Her friend followed by saying, 'If I do sweet and sour chicken the kids won't eat the chicken, yet they love roast chicken, they love chicken shakers'.

For one participant there was a certain irony in her children's love of chicken: 'We probably used to eat more of it than I do now, but that's because of the children [getting older]. I probably used to have chicken four or five times a week — probably a contributing factor in the breakup of my marriage 'cos my husband hated chicken! I didn't realise it until he was walking out the door'.

Half of the workgroup with younger children said that chicken was the most important meat in their family's diets, with one female participant saying that she didn't want to eat more of it: 'We eat chicken about twice a week, chicken is fairly popular in the family, [but] I get a bit fed up with it, [and] would prefer more of a range'. She was the only female who voiced this concern, until people came to their anxieties about what chickens were fed. This topic is expanded upon later.

- Price

Almost all focus group participants stressed the connection between the amount of chicken they purchase and its low price, relative to other meats and fish. In general, respondents said that they would not consistently pay more for chicken meat and they were in the main reluctant to buy more expensive free-range chickens. On being asked whether they would pay a dollar more for a free-range chicken, the following interchange from the lower middle class group was typical of the replies:

> STEVE: 'It's all about the free market. If one chicken is one dollar more than another, people will buy the cheaper one I guarantee you'.

CHRIS: 'We will buy the cheaper eggs not the free-range ones'.

KAREN: 'To be honest I would go for the cheaper one'. Everyone agrees.

SHIONA: 'If a chicken was one dollar more and advertised as much better I might start off that way. I probably wouldn't continue that way, not the week after, and the week after'.

JEAN: 'Organic food doesn't taste better, so I'd probably pay the normal price. Chicken — its just food'.

Male resistance to the alternative was present in one of the workgroups. 'I probably would [pay more for free-range chicken] but my husband who does most of the buying wouldn't pay the extra', said one. Another remarked, 'Trevor [an ex-butcher] wouldn't pay the extra'. And another husband was added to the list of sceptics: 'My husband doesn't believe in free-range chickens. He actually works for Franklins Fresh[11] and he has these conversations with the butchers and they say [free-range] is not true — with the eggs as well as the chickens'.

- The white meat alternative to red meat

For health reasons and for the purposes of pleasing everyone, the playgroups and one of the workgroups seemed to stress the importance of alternating the types of meat. The remark: 'We alternate white meat and red meat, so we eat quite a lot of chicken and pork — not much fish' was typical. It is worth noting that a need to balance red and white meats was only mentioned by participants where adult males were a regular part of family life.

- Versatility and ease of preparation

One who regarded take away chicken as generally too greasy was, however, impressed by the meat's overall versatility: 'I don't think a week would go by where I don't cook chicken fillets in some way'. Chicken's ease of preparation and cooking seemed important in both workgroups. As one participant said, 'Chicken — it's easy; being lazy it's easy to whip up'.

- Easy to chew

In the workgroups, an associated attribute to the ease of preparation was added to chicken's popularity. One mother commented, 'Laziness on behalf of the toddler who won't chew much [is a problem], so breast chicken [is good] for small children'. Another explained, 'We have chicken several times a week in all different ways, 'cos we thought it was healthier as well as being so flexible with what you can do with it. The kids love it 'cos its easy to eat and bland'. One of the fathers applied this reasoning to himself: 'I must admit when I'm tired and having been working physically hard all day and [I] think [about] what will we have for dinner [on the weekend] and we get take away chicken and chips — just the way it falls off the bone [is] beautiful'.

- Important for vegetarians and would-be vegetarians

One participant had a vegetarian wife and he explained how 'chicken is very important 'cos its the only meat substitute ... yes, she likes chicken'. A female colleague followed with her current dilemma: 'I'd like to become a vegetarian and my partner would like to eat ten ton of red meat and the compromise that I suggested is say over the weekend we have one vegetarian main meal, one chicken and one fish main meal — so the chicken is an acceptable meat compromise that's not red meat and makes him feel nurtured'.

CHICKEN'S NEGATIVE ATTRIBUTES

As foreshadowed however, chicken evoked disgust from several participants. This was particularly so for one of the workgroups who had earlier expressed how important chicken was in the household diet. A number of this group's participants had backgrounds in health or were working in that area at the time of interview and this appeared to temper their views toward chicken. One explained that she had gone 'off chicken for about a year when I did OT [occupational therapy], 'cos I reckoned the meat of a cadaver looked like cooked chicken — it was gross'. With everyone nodding in agreement, one mother said, 'I'm a bit suspicious of chicken ... I hear all these stories about hormones and I look at the fat on it and it makes me feel sick'. The father who was quoted above as saying how much he enjoyed chicken, when tired, later said, 'I do eat a bit of it, there's a shop in Moonee Ponds. I reckon their chicken tastes prefabricated, poxy, revolting — tastes artificial, you have to have a beer with it to cut through the grease'. At this point the women chimed in with 'red wine'.

Less extreme responses were voiced by approximately half of all participants. The word *concern* is appropriate for the doubt surrounding their chicken consumption. The topic of the chicken's diet and its impact on human well-being caused some heated discussion in one of the workgroups. In reference to her family, one said, 'We're becoming more conscious and worried about what they're putting into chicken and we're eating less of it and I would like some reassurance. But I tend to buy corn-fed or free-range chicken from the market. [I am] increasingly becoming more vegetarian'. Later a more representative comment was made: 'You can buy a free-range chook but I'm a bit sceptical about how free-range it is — maybe they've fed them hormones. It's the artificial things that get added to their diets that concern me. As long as its death is humane then it doesn't concern me, but it's the additives to the diet that concern me'.

Animal welfare concerns were voiced in both middle SES groups, which led me to ask, 'Would you pay more if you could ensure that the chicken had a good life?'

'Yes, yes, absolutely, yes and not have things pumped into them, yes ... if it would taste better' was how one in this group initially responded to this question. They then qualified their response accordingly:

> PETER: 'It's not so much a matter of a good life and death, 'cos if you're only grown to be eaten it wouldn't matter, but from a health point of view ... I s'pose that a good life implies a healthy animal, yes'.
>
> ANITA: 'So, not a good life, but a healthy life. It's more than chemicals and the other stuff, it's the morality aspect as well as the taste'.

The interchange was similar in the other workgroup. As one remarked, 'It's difficult when it's an everyday thing, but in our household 'cos food is so important, I'd happily pay substantially more and go without other things', while another ventured, 'I'd probably spend a dollar more if I knew there weren't additives. This sounds awful but the additives worry me more than the quality of the chicken's life. Battery versus free-range — I don't care so much about giving the chicken the choice, but the additives tip me over the edge'.

Agreeing to pay more for free-range chickens is one of the few areas that distinguish the groups and it is not surprising that the middle income groups are so disposed. Ironically the major reason for so doing was to avoid ingesting hormones they inaccurately believe to be fed to chickens. These participants were concerned that by ingesting the meat one also consumed the hormones. Three people gave this as the reason for their declining chicken meat consumption. I had the task of explaining to each group that growth hormones for animals had been banned since 1959 in Australia. Instead, in the case of intensively reared chickens, they are fed antibiotics that hasten growth. This seemed a relief to many of the participants.

The great majority were resigned to the fact that all parts of the food supply suffered from potentially harmful additives and until they could be sure that other foods were free of health-harming substances they would continue to eat chicken. Their general reluctance to reflect on matters of animal welfare or the animal feed supply for fear of having to take alternative action was palpable. This response does not suggest consumers who feel a marked degree of authority over the food supply.

EXPERT AND MARKET RESEARCH REASONS FOR CHICKEN CONSUMPTION

The increasing consumption of chicken meat has attracted numerous studies to explain its success *vis a vis* other meats. In an early assessment, four reasons were given for increasing chicken meat consumption during the 1970s:

- a decrease in red meat due to education programs focussing on its high fat content
- an increasing number of take away chicken outlets
- changing costs of production and the mass production of frozen chickens
- availability of store and home freezers (Flint 1981, p. 31).

Australian food historian Michael Symons (1982) has lent his support to the last factor, arguing that chicken's success has been due to the advent of both the supermarket and home freezer. He has described in particular, how the supermarket freezer was accompanied by a packaging breakthrough in the early 1960s: Cry-o-Vac plastic film allowed consumers to view the frozen chicken, making the product more attractive.[12]

Flint's second factor, the growth in chicken take away outlets, is nominated regularly as underpinning chicken consumption growth in the 1970s and 1980s (Skurray & Newell 1993; Turner 1977, p. 69). Some specify that it was due to well-promoted fast food outlets, referring in the main to KFC (Larkin 1991).

Lower retail price and lower price elasticity, compared to other meats, have been cited most often as leading to the dramatic increases in chicken meat consumption (Larkin 1998). Figures from elsewhere show that chicken has exhibited lower than average price rises compared to the CPI (Lester 1996, p. 109). This has combined with a lower baseline retail price compared to other meats, as shown in Table 4.3, to encourage consumption increases from the late 1970s to early 1990s.

Table 4.3
Producer and retail prices for meat

Year	Beef		Lamb		Pork		Chicken	
	prod/retail*		prod/retail		prod/retail		prod/retail	
1978–79	100	367	100	294	100	330	100	176
1983–84	141	636	141	391	119	506	132	258
1988–89	182	861	181	531	164	650	169	312
1993–94	210	980	236	629	159	728	187	280

*The first figure is an index of producer prices, the second is the average retail price per kilogram in cents.

SOURCE ABARE 1994, pp. 21–22 and pp. 173–74.

The figures in Table 4.3 show that the demand for chicken meat is both income elastic and price elastic. These characteristics have assisted the industry in increasing its output and expanding its share of the fresh meat sector. Technology and scale economies have enabled the industry to hold costs down and gain a relative cost advantage over other meat products, with chicken retail prices declining in real terms. At the same time the real price received by the producers has been increasing at a sizeable rate. In comparison with the red meat sector, chicken producers are not as susceptible to price variations generated through climatic changes. Nor have they been subject, because Australia is a self-sufficient producer, to price fluctuations due to production cycles in exporting nations. The orderly marketing arrangements for chicken production described in Dixon and Burgess (1998) have meant a relatively stable price environment for consumers and for producers.

Supermarket poultry buyers have credited chicken meat processors with keeping abreast of socio-demographic and cultural trends. Similarly, Larkin, who has studied the industry indepthly, supports a cultural relevance argument. In one of his industry funded assessments he surmises that '[t]here has been continual changing emphasis for convenience foods, such as take away chicken, Chinese meals and TV dinners. Chicken meat has played an important role in satisfying these evolving community preferences' (Larkin 1991, sec. 2.2). Larkin's view that children appreciate chicken accords with the focus group research, but where his opinions depart from those of the focus groups is around chicken's status as 'a healthy, low fat product' (Larkin 1991, sec. 4.3). The best that could be said is that the focus group participants viewed chicken as healthier than the red meat alternatives. While the focus group research concurs with expert opinion and market research, it yields a more complex rationale for chicken's esteem. The research points to a regime of values underpinning chicken consumption, one that is concerned with a range of attributes rather than any single factor. Chicken is popular because it simultaneously satisfies a spectrum of needs in food, from family harmony and women's desire to nurture, to price, availability, health concerns, variety and convenience.

ORGANISED CONSUMER RESPONSES

I examined industry documents and consumer group newsletters for evidence of organised consumer involvement with, or resistance to, the chicken meat industry. This section is brief however, owing to the fact that very little has been done by consumer groups in relation to this particular industry. There would be much more to report if this was a study of egg layers.

Periodically, animal welfare groups have been vocal in relation to table chickens, and their representations in 1990 to the Senate Select Committee on Animal Welfare had some impact. The Committee reported two major concerns: the small space in which fully grown birds could move and consequential inability of the birds to exercise, and the bird-damaging transport systems between farm and processor. A lack of response by producers on the first issue has resulted in the national lobby group Eco-Consumer campaigning in its newsletters against the intensive growing conditions of chickens (Eco-Consumer 1996). The group has voiced fears about additives in the feed of birds grown elsewhere and has pointed to the overuse of antibiotics and hormones in other countries. The group struggles to keep financially afloat and no one I spoke to knew of its existence.

This is not to say that consumer group campaigns have not had an impact on the anxiety felt by the industry that it may be targeted by consumers at any time. The chicken meat industry's fears stem from the high profile media campaigns against the egg industry. Regular activist forays to release battery cage hens are well recognised and have been legitimised by the RSPCA,[13] which allows its name to appear on cartons containing eggs produced by better housed birds. This publicity caused many focus group participants to ask me about the growing conditions of table chickens. They were in the main surprised to hear that these birds grew in a cage-free environment.

How does a lack of organised consumer response accord with what has been previously written about consumer activism in regard to food? E.P. Thompson, who has documented the operation of norms and patterns of exchange regarding food in times of 'dearth' in the 18th and 19th centuries, uses the concept of the moral economy to describe his findings. He explains the 'moral economy' as exchange justified on the basis of social or moral sanctions, as opposed to the operation of market forces (Thompson 1968). In support of his argument, Thompson describes numerous food riots. He shows how collective action stirred by strong emotions associated with life's necessities, coupled with a sense of entitlement to such necessities, builds a sphere of life around the idea that the well-being of the community should come before market forces and the profits of a few individuals (Thompson 1993, p. 336–339).[14]

In a context of relative food supply abundance, my research shows little food related consumer agitation. Instead, the focus groups wanted a food supply that would help them deliver family harmony and ensure family member's health. The evidence from my focus groups confirm the findings of other studies that women covet food systems that allow them to continue the gendered duty of caring work. Women's desire to be responsible for constructing family life, based around food and home, found by both Duruz (1994) in

Australia and DeVault (1991) in the United States, activates only limited demands on outside agencies for assistance. And where such demands are made they continue to be for *pure*, or unadulterated, food. This does not negate the operation of a modern moral economy. Miller (1995), the architect of the housewife as global dictator, describes the moral economy of the household as built upon a more inward looking, duty-to-care basis than Thompson's concept. Miller's moral economy denotes a private realm of exchange, meaning creation and consumption that are in contrast to making ethically founded demands of the marketplace. In this sense, chicken meat consumers appear to be reproducing a sphere of life that stands outside the market economy, but one that is less demanding of commercial firms and governments than one might expect. This finding begs a further question: how are producers, retailers and other food experts reacting to and influencing moral economy concerns? This theme continues in the successive chapters and is bound up with the question as to the role of authority relations in the balance of power between consumers and producers.

POWERFUL CONSUMERS OR POWERFUL TASTES?

Whilst limited in number, the focus group interviews confirm market research and expert opinion that chicken is popular for no one reason. Chicken appears to have been important in all of the households at one stage or another, but for reasons that vary over time and between households. In particular, the groups revealed that parents appreciate chicken because it is economical, easy to use, readily available and relatively non-contentious in terms of family harmony. Most importantly, chicken appears to be a good compromise food or, as one food sociologist has put it, the least controversial of all meats (Whit 1995, p. 9). This conclusion was supported by chicken's inclusion in at least one vegetarian diet.

Chicken is not only a good compromise food for non-meat eaters; it accommodates what, on the surface, seem to be other contradictory qualities. It is appreciated by health conscious individuals who also like fried take away food and it appeals to those who are busy but are prepared to take the time to cook a whole bird. Chicken eases the burden of family cooks and food providers by its ability to please as many family members as possible, perhaps more effortlessly than other meats. In these ways, chicken meat makes a substantial contribution to a relaxed state of being which, as seen from Chapter 1, seems a desirable attainment for many Australians.

The significance of chicken's low price cannot be totally dismissed. Rather, price was a key contingent in its popularity, especially for those in the low SES groups, and it could be assumed that its consumption rate would diminish if the price grew relative to other meats. Possibly this is due to the reservations that many have towards its preparation and production processes. While focus group participants described chicken's many positive attributes, they invariably qualified their statements with ones which showed that thinking about, preparing and eating chicken caused them disgust, disquiet and concern — the major one being the additives in the birds' feed. Amongst the additives mentioned consistently were hormones and this reflects the more general ignorance that participants had of the primary production process. However, without exception all wished to remain ignorant of, as one participant put it, 'the true story' of chicken production because of the importance of chicken in household diets. 'I can't afford to know', as another participant explained, sums up the prevailing sentiment. The focus group responses provide ample evidence that being concerned about a food is not sufficient to stop consumption if other incentives are present. In addition, the discussions confirm how potent the messages about fat in food have been. Throughout the 1980s nutrition scientists and parts of the medical establishment labelled red meat as not contributing to *good nutrition* (Crotty 1995). The fact that the fat content associated with red meat was perceived by the focus groups as more harmful than the fat content of chicken meat illustrates the success of that particular campaign.

In short, chicken is a popular food because of the range of values it represents as opposed to any particular hierarchy of values. Furthermore, perhaps owing to an absence of cosmopolitans in the focus groups, I did not glean differential assignation of meanings or of behaviours. The socio-economic dimension to chicken meat consumption appeared only when participants were asked if they would consider buying more expensive, chemical free birds. In every other area, chicken appeared to be imbued with similar cultural and economic meanings. The values that consumers attach to chicken meat and its consumption are singularly prosaic and could not be expected to enhance the status of the shopper, meal preparer or eater. Because chicken meat does not currently provide a vehicle for status distinction, it can be interpreted as a relaxed food. Any claims about consumer empowerment through their attachment of values not intended by the market appear not to extend to chicken meat.

So what about the claims made by agrifood sociologists that new flexible production systems have emerged as a response to consumer demand? On the evidence so far, consumer influence over chicken meat production systems appears to be negligible. While consumers obviously appreciate the transition from the one-size-fits-all, frozen chicken to the seemingly endless variety of chicken meat products, there is no evidence of consumer pressure on producers to address their real concerns: less fatty and chemical free birds. Furthermore, a mix of cynicism and extra cost precludes a true niche product — free-range or organic birds — being considered an alternative. Indeed, cynicism about who and what to believe reinforced a feeling that being better informed about production methods would not help, because consumers cannot believe what they are told. Gofton (1990) and Beardsworth and Keil (1997) attribute a cynical attitude to a combination of factors, namely the changing context in which less food is cooked at home, young women's ignorance of fresh food, greater consumption away from the home, and mass media encouragement of greater experimentation. Thus, while the post-Fordist attribute of product heterogeneity is present, the other post-Fordist attributes of quality and basis in lifestyle niches are difficult to discern. The focus groups are consuming a product that comes in various mass produced shapes and sizes, but which disappoints in terms of quality. Possibly due to the limited range of socio-economic status represented in the focus groups, there was no support for the suggestion that consumers are rejecting Fordist products.

CONCLUSION

The story of chicken meat in the Australian diet as told in this chapter is not a story of powerful consumers but of a powerfully ambivalent consumer taste for a food that is extremely important in family diets. As the backyard production of chicken disappears and is replaced by ignorance of industrial production, consumer power over this part of the food system appears negligible. The research shows that consumers are uncertain about what constitutes good food, and while they take its supply for granted they cannot assume food's safety, nutritional worth or a universally pleasing taste. For the consumers consulted for this research, decisions about what is *good food* are arduous and pose a constant dilemma for the family food provider. In the absence of organised consumer demand for different chicken meat products the authority rests at this point in time overwhelmingly with producers, retailers and health professionals.

The fact that Australian consumption of chicken is growing in line with other parts of the world highlights a need to explain the production of tastes for foods and food practices. A strong case has been made in the literature that taste preferences are acquired through multiple pathways and that a multi-level set of principles operate to move food through being *good to think* to being *good to eat* and finally to the stage of being *good to taste* (Falk 1991, p. 758). In the Australian context of relative food abundance and with rampant reflexivity around dietary matters, it is timely to highlight the process of taste production. The role of post-war producers and retailers in a generation with a taste for chicken is the subject of the next two chapters. The material that follows examines the extent to which producers feel pressure from consumers to produce niche products and to yield to considerations such as quality and lifestyle deliberations.

5

PRODUCING CHICKEN: WORKING WITH REAL TIME

The birth — or more correctly — the hatching and the growth of the chicken meat industry in the early 1950s to its present size and standing as one of the major protein food sources for the Australian population is perhaps the greatest success story of development of any primary industry in Australia.

(Bell cited in Cain 1990, p. 9)

Earlier material creates the impression that poultry industries are highly successful because they are at the forefront of post-Fordist production regimes, at least within the agricultural sector. Poultry, particularly chicken meat, industries are claimed to be introducing Just In Time procurement systems, global sourcing of inputs and integrated producer-grower relationships in advance of other agrifood commodity complexes (Boyd & Watts 1997). Poultry production's efficiency is attributed to these features. Moreover, it is said that poultry industries are providing for niche markets built upon concerns for quality and lifestyle through the adoption of flexible production methods. This particular reading of the Australian chicken meat industry receives little support from the previous chapter where we witnessed marked consumer indifference to anything other than what Fordism has to offer. Much more can be learned about the existence of post-Fordism by examining the product development process, to ascertain the extent to which product differentiation results from consumer demand. This is a feature of too few commodity analyses.

This chapter is based on interviews with primary and secondary producers, their respective associations, site visits, attendance at relevant conferences and reports from the chicken meat industry and from government bodies. The material has been organised according to the production sphere headings of the amended commodity analysis framework outlined in Table 3.3. The headings are re-ordered to facilitate a better flow in the narrative. My analysis of 'regulatory politics' appears in the final chapter due to its relevance to understanding the emergence of the global chicken. What follows here is an overview of the production side of the Australian complex, including free-range producers. A chicken's life is described, as are the working lives of those who are most intimately connected with that life and death, namely the hatcherymen, farmers and process workers.

CHICKEN MEAT PRODUCTION OVERVIEW

No government records were kept until the mid-1960s, but industry sources estimate that in 1950–51 Australians were producing three million broilers, with production growing seven fold in the fifties and five fold in the sixties.[1] In the 1970s production doubled and the last fifteen years has seen further steady growth culminating in 329 million chickens being slaughtered in 1996. Indisputably, Australians have shifted their protein intake from two grazing animals — cattle and sheep — to a factory animal and in so doing have laid the basis for transforming the culinary culture.

An overview of the industry reveals approximately ninety processors and all are Australian owned companies. Until 1999, the industry described itself as three tiered, with Inghams and Steggles — known in the industry as the Big Two — dominating about a dozen medium-sized processors and seventy or so smaller producers. The industry remains three tiered, but the Big Two comprise Inghams and Bartter, the latter having bought Steggles from Goodman Fielder for $131 million in 2000.[2] Until the acquisition, Bartter was the largest of the medium sized companies.

At the bird rearing stage there are some 820 growers, generally self-employed family businesses, who work under contract for one or other of the large and medium processors. The largest processors also employ people to work their company farms. The farm labour on the corporate and family farms that breed, hatch and raise the birds amounts to about 13 000 jobs with a further 7000 people producing feed grains for the industry (Fairbrother 2001).

The capacity to mass-produce chicken meat was consolidated by two of the long established poultry families, Inghams and Steggles, adopting what is known as the vertical integration industry model thirty years ago. This meant that a single company sought to own chicken breeding and hatching operations, feed mills and processing plants, and to augment its own growing operations by contracting out the rearing of chickens to other farmers. The order imposed by vertical integration was such that, unlike in the egg industry, the powerful players in the chicken meat industry did not need to have a marketing board to align sales and production. Figure 5.1 lays out the vertically integrated production side of the chicken meat complex.

Figure 5.1
Chicken meat production

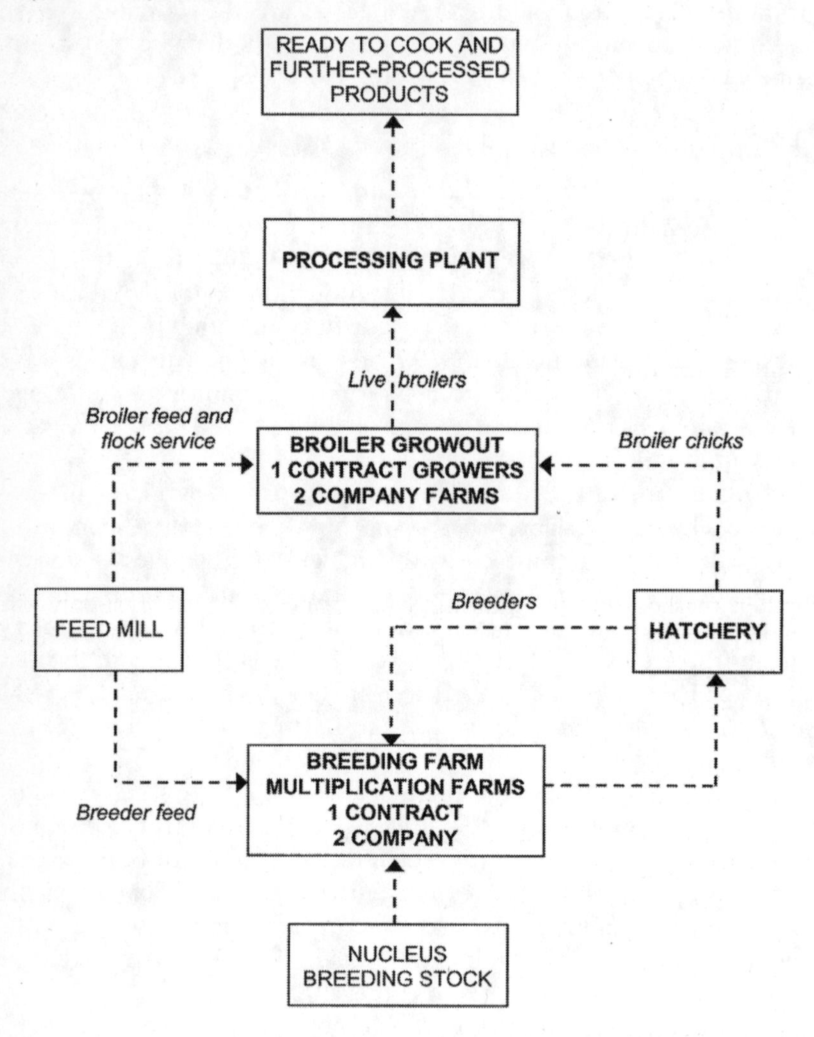

The largest processors have their own nucleus breeding stock, breeding farms, hatcheries, company farms that grow the birds from day old to slaughter weight and processing plants in each state, in addition to owning the birds that the contract growers raise. Until buying Steggles, Bartter of Griffith, NSW, was Australia's only truly integrated operator. It had its own grain farms and was doing all of its own poultry growing. In other words, a single company can be involved in every facet of primary and secondary chicken meat production. This is vertical integration at its highest level and few other agrifood complexes come close to achieving this.

Even with the increase in bird production the number of processors and farmers has remained fairly constant over the last decade and despite new technology the number of process workers has increased. Approximately 40 000 people are employed in the processing operations and unlike the beef, sheep meat and pork industries, the poultry industry has been increasing its share of total meat production since 1981. In 2000 there were 120 000 Australians directly and indirectly employed to produce and sell chicken meat (Fairbrother 2001).

Four major industry bodies dominated the industry in the 1990s: the Australian Chicken Growers Council (ACGC) which represents the poultry farmers; the Australian Poultry Industries Association (APIA) representing the two largest processors; the National Poultry Association in which the dozen or so medium sized processors assemble on a 'needs-to' basis; and the Australian Chicken Meat Federation, an umbrella organisation for all players on the production side of the industry.

CHICKEN MEAT INDUSTRY OVERVIEW

PROCESSOR CONCENTRATION

In 1985 the Prices Surveillance Authority (PSA) was asked to examine whether market power resulting from concentration of supply of chicken stock was being reflected in prices charged to customers and consumers. The PSA found a predatory environment had been operating between 1968 and 1980. The major players — Inghams, Steggles and Inghams/Amatil (British Tobacco) — were buying out small independent state based companies around Australia, successively increasing market concentration. By controlling the supply of day old chickens and bird processing, the large companies were in a position to put a cost and price squeeze on poultry farmers. This situation was exacerbated in 1980 when the Steggles family sold to Amatil, leading to a near monopoly arrangement.

Not surprisingly, the PSA expressed concern about the lack of competition in the supply of day old chicks, which at the time of their enquiry, was effectively controlled by Inghams. Furthermore, the PSA was not satisfied 'that major processors related by ownership were operating at arm's length in marketing dressed chicken' (PSA 1986, p. 2). As a result, the PSA indicated its intention to oversee future bird price rise decisions by the processors. A year later Amatil began to sell its poultry businesses, culminating in its sale of Steggles to Goodman Fielder in 1989.

Tables 5.1 and 5.2 show the dramatic decrease in the number of processors over the last twenty years for South Australia. The picture for this state is by no means unique (Cain 1996).

Table 5.1
Table chicken production in South Australia, 1974

Processor	Total	%
Windsor	83 000	27.7
Manos	70 000	23.3
Inghams	55 000	18.3
Pape	40 000	13.3
Aidon	25 000	8.3
Goldalla	10 000	3.3
Norlunga	6 000	2.0
Mac's Chicken	3 000	1.0
Tower Poultry	3 000	1.0
Baradakis	2 500	0.8
SA Poultry	1 500	0.5
Other	1 000	0.5
Total	300 000	100

Table 5.2
Table chicken production in South Australia, 1995

Processor	%	Combined total
Inghams	98	
Steggles		
		552 000 per week
Joe's Poultry	2	
Others		

SOURCE Cain (1996).

In 1982–83 the top four enterprises controlled sixty-six per cent of industry turnover (PSA 1986). Industry concentration figures for 1996 show that the three largest processors account for seventy-six per cent of industry turnover, and seventy-eight per cent of chicken meat production (Table 5.3). This compares with a concentration level of forty per cent by seven firms for the other intensively produced meat, pork (Plunkett et al. 1996). Concentration among chicken meat producers, it seems, has intensified due to the departure of a number of multinational feed and food companies arguably hastened by the PSA's decision to monitor competition between processors. Moreover, further concentration has occurred in spite of a trade liberalisation decision in 1989 to open the Torrens Island Quarantine Station for the processing of imported avian stock. This decision reversed a forty-year ban on such stock and guaranteed smaller processors access to birds without relying on Inghams and Steggles for their supply. It was hoped that the decision would introduce competition into the industry by encouraging the expansion of the medium size processors.

Table 5.3
Chicken meat industry concentration, 1996

Producer	Chicken meat production (%)	Chickens per week (millions)	Employees ('000)
Inghams	40	3.0	6.0
Steggles	28	1.8	2.7
Bartter	10	0.5	1.1
Total market share of these three (%)	78	76	89

SOURCE Table compiled on the basis of data supplied by the Australian Chicken Growers Council (1997).

Of the largest processors, Inghams produced 3 million birds a week in 1997, followed by Steggles at 1.8 million and Bartter at 520 000. The Australian industry is far more concentrated than the American industry where, in 1994, the four biggest processors had forty-two per cent of the market (Boyd & Watts 1997, p. 201). Other figures at that time suggest that Inghams and Steggles provided up to ninety per cent of meat in one state and never below half of any state's production (Dixon & Burgess 1998).

It is fair to assume that the statistics in Table 5.3, which preceed the 2000 Bartter take-over of Steggles, dramatically understate the industry concentration. Despite the prevailing concentration situation, the Federal Government's Australian Competition and Consumer Commission (ACCC) ruled not to intervene on the sale. In its media release the Commission stated that '[c]ompetitively priced chicken is important for households and businesses as it is sold in many supermarkets, fast food shops and restaurants. During inquiries the ACCC spoke to a number of growers, processors and major customers. These discussions indicated that the acquisition would be unlikely to result in a significant reduction in competition in the industry' (ACCC Media Release 24 August 1999).

GROWER GROUP ORGANISATION

The official history of the industry written by Cain (1990) describes how the rapid creation of a mass market, made possible by the introduction of imported chain processing equipment and the proliferation of bird breeding companies, encouraged retail price discounting wars in the late 1960s. Inghams, Steggles and at least one smaller operator in each of the states battled to obtain market share by offering the cheapest chickens possible. Frozen chickens became 'loss leaders' for the supermarkets: low prices for frozen chickens were a means of encouraging consumers through the supermarket doors. To achieve this the processors had to be ruthless with the inputs they applied to their business, and the input over which they felt they had most control was the contract growing fee paid to the farmer. The processors began to offer successively lower fees leading to situations where farmers refused to take birds onto their properties. Cain's history reveals the intensity of the periodic disputes between growers and processors and how, by the end of the 1960s, state governments had intervened to mediate the disputes. New South Wales, then all states except Tasmania, introduced legislation to oversee the contractual arrangements between farmers and processors.

In the early 1960s the first grower's association was set up in New South Wales. Within a few years each state had such a body concerned with similar issues, specifically the growing fee paid to farmers by processors. In one of its first acts, the NSW Broiler Growers Association fought to replace a system under which it 'was not unknown for a grower to be presented with a bill for the privilege of growing a batch of chickens' (Cain 1990, p. 73). Instead they advocated a scheme whereby growers were paid for raising birds. This was a time when Inghams and Steggles made use of their company farms whilst denying contract growers any day old birds.

In 1968, the NSW State Government was forced to intervene and at a meeting chaired by the Minister for Agriculture, all parties agreed to arrangements to introduce orderly marketing of birds and to ensure a fair growing fee per bird was paid to growers (Cain 1990, p. 73). Within a year prices had slipped back markedly for some growers and disagreements continued up until 1975 when the NSW Minister agreed to introduce legislation to cement the arrangements. Industrial action flared again in 1983 and 1986, leading to a new Act in a further attempt to improve matters.

Similar events were taking place in Victoria. In 1969, Victorian growers went on strike and refused to grow birds for seventeen weeks. Growers were eventually forced to accept day old birds because of the prevailing monopsony arrangements which dictated that the growers had no alternative sources of bird supply. The 'unease' continued for four to five years:

> when, after a second strike of 16 weeks, the Minister for Agriculture appointed an arbitrator, and in 1975, the Victorian Broiler Industry Chicken Industry Act was passed through Parliament and proclaimed January 1 1976. The Bill established the Victorian Broiler Industry Negotiating Committee (VBINC) and made it illegal for anyone to grow more than 500 chickens at any one time without a contract approved by the Committee (Cain 1990, p. 91).

The man credited with organising chicken farmers, first in Victoria and then nationally, was Wally Shaw, past president of the Victorian Farmers Federation. Forty years ago, Wally and his wife Wendy, both agricultural science graduates, were struggling financially with a few dairy cows. Around 1960, the Shaws expanded by adding some broilers, which led them to meet regularly with other farmers to exchange ideas about equipment and growing techniques. Within a few years processors were requesting that the Shaws grow their birds. According to Shaw, he and his neighbours found themselves being offered growing fees that actually declined over time. Alarmed at this, as well as the degree of intimidation being experienced by 'New Australian' farmers,[3] Wally toured the state encouraging chicken farmers to form groups and to join the Australian Primary Producers Union.

Shaw remembers the 1970s as 'a very bad time', especially due to the presence of two particular corporations — British Tobacco and Canada Packers. The farmers wanted contracts with a set fee and in order to achieve this they refused to accept birds onto their properties, and went for periods of six weeks and more without any income. They held out because, as he put it:

> We had comradeship, because we were targeted. And it helped to be targeted by multi-national companies. Day after day we would get forty to fifty farmers to blockade properties. One time about 4 am twelve processor trucks lined up and the police came. 'Let in the trucks', the police said, 'but we won't necessarily show up again'.

In the first arbitration following the 1976 legislation, the price was increased by twenty-five per cent per bird. Shaw believes that contract farming has advantages: it is 'super' efficient; guarantees farmers an income; and helps spread risk. 'The downside is that unless the farmers are organised they can be intimidated'. He also believes that the way the collective ethos is managed is very important: 'one vote one farm, no matter how big the farm or number of birds'. Under Shaw's leadership of the state farmer's federation, this principle became two votes per farm to recognise the contribution of wives, sons or daughters and siblings who together ran farms. His other guiding principle is 'no matter how big you are you always stick by the little fella', and 'you have to show that you cannot be bought or frightened'.

There are local grower organisations in each chicken-producing state which are affiliated with the Australian Chicken Growers Council. The pivotal role that state and national grower's groups have played in fighting the deregulation of the industry is described in Dixon and Burgess (1998).

OUTSIDE THE VERTICALLY INTEGRATED, MASS PRODUCTION COMPLEX

Little is known about the free-range chicken meat industry. However, I have heard it admitted in conversation with poultry growers and processors that chicken labelled free-range is often not. Confusing the issue further is cynicism over whether free-range birds exist at all, a concern voiced by numerous focus group participants.

In the mid-1990s, it was estimated that of the 171 000 birds entering the Victorian market each week, only 1000 were free-range. Presently, only one per cent of chickens consumed in Australia are free-range, supplied by poulterers who are generally integrated growers/processors/distributors and who source their own feed from neighbouring farms. Two free-range farmers I spoke to did not identify with the poultry industry: rather they saw themselves as part of the game industry. The day old chicks used in free-range farming are procured from the major processors, with chick price and ability to supply determining where the free-range farmers source their birds. These producers generally grow and slaughter the birds on the same property and most owner-operators distribute the dressed birds rather than rely on a marketing agent. Free-range production methods are described later in the section headed 'Production practices'.

SCIENCE PRODUCTION AND APPLICATION

Science has been critical to the success of the chicken meat complex and it is alleged that researchers know more about the genetics of chickens than about any other domestic animal, humans included (Boyd & Watts 1997, p. 193). Indeed Cain (1990) attributes this success to 'men's' interest in breeding and hatching birds. As in North America, reliable mass hatchings of 'meat' producing avian stock rapidly transformed what was a labour intensive, cottage based system to a Fordist, mass production, capital intensive system. This development allowed poultry farmers to diversify from egg production and to profitably engage in chicken meat production.

A case in point is Bert Tegel who, working from his father's cow, pig and poultry farm, designed and built incubators and hatchers (Cain 1990). Young Tegel went on to study and gain his bird sexing certificate in 1934. The cloacal method of bird sexing enabled the separation of sexed birds, critical in egg production but also important in the early days of rearing table chickens. Over the next twenty years Tegel continued researching chicken meat strains and in 1952, after one of many trips to the United States, commenced a breeding program to improve commercial laying strains. There was a certain urgency to this work in light of an Australian Government ban on the importation of avian stock. Tegel's work bore fruit when he introduced the TM1, Australia's first meat chicken, at a field day in 1959, followed a year later with the improved TM4 bird. The Tegel company entered into franchise agreements to supply birds to a handful of firms across Australia including Inghams in New South Wales and two New Zealand operations. In 1963, the Ingham brothers bought a half share of the Tegel business: a partnership that continued until 1987 when Tegel became a subsidiary of Inghams Enterprises.

By the late 1950s, other entrepreneurs joined the breeders and hatchery workers and the few pioneering poultry farmers to create an industry which would over the next decade replace hundreds of independent, self-contained backyard breeders, growers and processors. It is little wonder that the Australian poultry industry has been declared 'technology's child' (Fairbrother 1988) and the role bird breeders have played in the Australian context is consistently highlighted. According to Blackett:

> [u]nlike the United States, Canada and the United Kingdom, the Australian poultry industry has taken its bearing or pivot from the breeders of the parent stock. Feed companies in the overseas countries largely inherited a bankrupt business in the fifties and thus became the financial hub of the industry. In retrospect the way our industry has grown has been a much happier experience for all concerned (Blackett 1970, p. 447).

After genetics and avian management, the Research and Development (R & D) priorities shift to the economics within processing. At the biannual Agricultural Outlook Conference in 1996, delegates discussed the possible monitoring of overseas developments in gas stunning and new techniques for handling birds (Fairbrother 1996, p. 246). The production of special feed grains as a cost cutting measure is also of importance, given the limited control the industry has over feed costs. Effort is currently centred on growing wheats suitable for animals (as opposed to humans).

Perhaps just as significant as recent scientific innovations is the increasing use of scientific argument and discourse in the policy arena. The way in which risk assessment concepts are being used by many players to argue for and against the global free trade of chicken meat products and birds provides ample evidence of this assertion (Dixon & Burgess 1998).

PRODUCTION PRACTICES AND THE LABOUR PROCESS: PRIMARY PRODUCTION

Nucleus stock farms provide the sites in which the scientific discoveries, just described, currently take place. They operate like scientific laboratories, breeding into the grandparent stock qualities such as disease resistance, efficient meat to feed conversion ratios and survivability. Fifteen years ago the average weight of a bird hovered around 1.8 kilograms, now it is 2.2 kilograms. Some farmers worry that while the birds have become brilliant feed converters, the meat has lost its taste. Others contend that structural deformities in birds may be attributed to the speed of their growth.

As Figure 5.1 indicates, the nucleus stock moves onto breeding and multiplication farms from whence eggs are transferred to hatcheries. Day old chicks then move onto chicken (broiler) growing farms.

It is in these latter sites that the primary production processes of the industry take place.

THE BREEDING FARM AND HATCHERY

The nucleus stock farms supply breeder farms with males and females to mate — generally one male to seven to ten females, who, at twenty-four weeks, will begin producing fertile eggs. The birds lay until seventy weeks of age, each producing 150 to 160 eggs in their lifetime. One breeding complex may be worth $4 million, and some argue that they are the invisible key to successful chicken meat, for it is there that the scientists' endeavours are taken to the next vital step.

The eggs are transported from the stock farms to the hatcheries, which incubate the eggs over the next twenty-one days. Most hatcheries accept eggs from at least two sources to ensure a steady supply of eggs, or in the words of one hatchery worker, to ensure 'we're not stuffed'. A tour of the Fiveways hatchery in Victoria gives an insight into the early days of the chicken and of the workplace stress involved for workers.

The manager of this particular hatchery, owned by Eatmore Poultry, lives on the property, which consists of a number of unprepossessing buildings with extraordinarily sophisticated machinery. The Canadian firm, Janeway, supplies much of the equipment. If any mishaps occur the hatchery manager can access the firm's mainframe in the United States, and via satellite link make a diagnosis of the fault and obtain directions for rectifying the problem. Computers talk to computers in the task of turning embryos into chicks. The pressure on all staff is palpable. Issues such as temperature control are critical: on this site the manager has only eight minutes between a serious malfunction and the chicks dying. His house is wired to the plant, and he is wired to the property via mobile phone. This is a twenty-four hour a day responsibility.

When the eggs arrive they are dated and assigned a place in a cool store. There the embryos are turned automatically on the hour to 'exercise' them. The manager insists to his nineteen, mainly casual, staff that 'we deal in embryos, not eggs. You buy eggs at the supermarket. You have to take every care of them here'. The embryos stay in the cool state for up to seven days, after which they are moved to one of about thirty heated chambers, or incubators, where they sit in racks for about eighteen days. After a further three to four days they hatch. On any one day, out peek tens of thousands of pale yellow heads. At Fiveways, a relatively large hatchery, 430 000 chicks are delivered per week.

A handful of women are handed trays of eggs in various stages of hatching and their job is to help free the chicks of shells. They rapidly dispatch the sickly ones to a box on the floor — a one and half per cent reject rate — and put handfuls of healthy chicks onto a conveyor belt which is about twelve feet long. As they come down the belt a fine mist of vaccine for bronchitis covers them, entering through the eye. This all happens between 8 and 9.30 in the morning. The vaccinated birds then drop in an undignified manner, like soft tatts-lotto balls, into waiting plastic crates that move slowly along another conveyor belt. The crates are stacked ready to be put onto the company trucks and driven out to company owned or contracted farms, or possibly to farms aligned with other smaller, independent processors. Up to 100 000 chicks will leave the hatchery by 11 am and will arrive on the farms two to three hours later.

GROWING THE CHICKENS

The day old chicks are delivered onto the farm by truck, quite possibly one painted in yellow and blue nursery colours, over two to three days, after which they are placed into a cavernous, cage-less shed, characteristically situated on a small farm of between twenty and fifty hectares. The farms are generally clustered around a processing plant, and are located within close proximity to major cities. Location was determined some twenty to thirty years ago by the economics of transport in relation to markets, labour and processing facilities. Availability of reliable water and electricity supply has also had an impact on location choice.

In Victoria an average farm tends 60 000 birds at any one time, five to six times a year, thus producing between 300 000 and 400 000 birds annually. A medium sized processor needs an arrangement with up to thirty farms to supply sufficient birds for its annual operation. This is what happens on any one day on a Mornington Peninsula farm which supplies Victoria's largest medium sized processor.

Located approximately seventy kilometres from the centre of Melbourne, the property is set in beautiful surroundings, with chicken growing sheds and feed silos a short walk from the modern farmhouse. Prior to the delivery of chicks the farm owner, Rod, has cleaned up after the preceding batch of chickens. Another truck has visited and removed the old litter. He has washed out the sheds, sanitised them and ideally let them rest, as well as having a bit of a rest himself. The break between batches varies between one and three weeks, depending on the processor's demand for chickens. When the call comes that the chicks are about to arrive Rod lays the sheds with fresh litter of rice hulls because they are, in his opinion, very absorbent and relatively dust free. Other farmers prefer a bedding of composted old broiler litter. He does the last minute checks of equipment — feed lines and water feeders — and makes sure the heaters are working. Indeed, the chicken housing is pretty fancy and there is very little chicken wire in sight.

By the time the chicks arrive the first delivery of twenty-four tonnes of feed has been made. It will consist of a mix of wheat, sorghum, oats, field peas, lupins, soybean meal, canola meal, other protein, minerals and premixes. The feed grains are primary products traded as commodities and as such are subject to price fluctuations resulting from export and domestic market conditions. The object of poultry feed formulation — given that it is the highest cost in chicken production — is to achieve the lowest unit cost of bird production (Larkin & Heilbron 1997). I was told by one industry leader that because Australian chicken farmers have never enjoyed feed subsidies, the industry has produced the best feed converters in the world.[4] The feed contains antibiotic growth promotants, as allowed for in veterinary protocols, in the early weeks of the birds' life. Antibiotic growth promotants are not the same as the growth hormones so feared by consumers, but their use does alarm some public health figures (Chapman 1999). The price of each ingredient, coupled with the particular bird stock's requirements, will dictate the precise ingredient mix.

Processor-employed vets, charged with formulating the feed, will know exactly what can be spent on this part of the process and it is their job to ensure that the 70 000 'new souls' that have just arrived will lay down as much meat as possible during their stay of five to eight weeks. The Australian industry expects that birds will average conversion ratios of better than 500 grams in carcass weight for every one kilogram of feed grain consumed over the bird's lifetime.

Initially the birds do not eat much and Rod will tend to them at least three times a day. In addition, his farmhouse is wired with alarms to tell of any major equipment malfunction. The morning work consists of walking through the sheds slowly, to make sure the temperature is okay because the birds are very temperature sensitive, and to ensure that no drinker leads are leaking because they are also moisture sensitive. The farmer is alert to the noise from the birds and their activity levels. Active birds are a sign of healthy birds; dead birds are collected and mortalities recorded. The height of the drinkers and feed lines are adjusted. As Rod pointed out '[t]he whole business of growing chickens is minimizing the stress on the birds — and this way you can minimise disease. Chickens are susceptible to disease because they are intensively farmed'. The farmer weighs the birds every seven days. This will indicate a normal digestive function and flock health. Once a week, if the farmer is contracted to a 'responsible processor', the processor's vet will visit to check on the birds.

The morning shed work will take between one and a half to four hours. Rod returns after lunch and late afternoon to check that everything is okay and then, before bed, he goes to the sheds once more. These visits may be as brief as half an hour. Each week, the maintenance tasks — 'there's a lot of gear down there' — will take several more hours. I was told that chicken growers do not enjoy summer. 'For the last three to four weeks chickens are very susceptible to heat loss, so you may spend all day, and up to midnight in the sheds, adjusting the various cooling systems: weighing up humidity and air flow. You occasionally wander off and jump in the pool'. Moreover, the work can be very dusty and in the United States they have isolated a condition which they call chicken farmer's lung. In Australia, farmers are responsible for wearing protective masks when entering the sheds, the purchase price of which is built into the grower's fee.

According to the Australian and New Zealand Federation of Animal Societies (ANZFAS), chickens have a space comfort zone. In its submission to the Senate Select Committee on Animal Welfare, ANZFAS recommended that stocking rates be reduced by a fifth in many sheds. The Senate Committee expressed its own reservations about the practice of twenty fully grown, two kilogram birds, sharing a square meter: amounting to one fully grown bird having less than an A4 page in which to live. In such crowded conditions, birds are unable to dust, bathe and scratch in the litter and these issues are of concern to some farmers.[5] Two of the four recommendations made by the Senate Committee for the broiler industry concerned stocking densities. There is no evidence that these particular recommendations have been acted upon.

After five weeks the farmer's busiest time begins with the 'pick-ups' of smaller birds destined for supermarket rotisserie sections and fast food outlets. The trucks arrive after dark when the birds are quiet, and prior to their arrival the farmer has spent half an hour raising equipment so that the bird catchers have easy access to the birds. It will take up to an hour to fill a semi trailer and then a further hour to reset the equipment. This will go on each night until all the birds, small and large, are removed. Farmers hope for young bird pick-ups because with each passing day the birds consume more feed and farmers are paid per bird not per kilogram.

Ideally, under contract to one processor for three years, this cycle will be repeated five to six times a year. For each batch the farmer is on call seven days a week, able to leave the farm for several hours at a time only. There are routines, but no days are typical because of the vagaries of the weather and equipment fallibility.

Rod is also a grower representative, meaning that he is a member of the processor committee that makes decisions about the grower payments. He spends a lot of time on the phone to other growers in his group checking on their progress. He sums up their collective lot in the following manner:

> ... it's not a bad life. Chicken growers in this state have done pretty well. Although you have to be able to handle stress — you've got to

> be able to handle heaps of stress. You've got a live animal, a lot of risk, hard work, but it's unglamorous. That's where you get the saying 'Oh, he's just a chicken farmer'. So it doesn't pay to take it too seriously, [you've] got to look outside your square, and yes, some do have fun if they can handle the stress.

In this statement the idea of risk management within vertical integration is manifest. However, it must be placed in the context of state regulated contract farming which arguably makes the risk sharing more equal than in deregulated poultry industries. For Australia's contract chicken growers, the farmer's personal investment and dedicated operation are impressive but so are their financial rewards. Contact with numerous farmers confirms the professionalism of the chicken growers, their specialist knowledge about bird genetics and behaviour and their pride in doing a good job. They may not own the birds but they see themselves as stewards of the land and of animals as surely as those who own and raise the larger, more prestigious cattle and sheep.

PRODUCTION PROCESSES AND THE LABOUR PROCESS: SECONDARY PRODUCTION OR THE PROCESSING OF CHICKENS

In the truck ride to the processor, generally not more than two hours away from the farm, the chance of birds becoming stressed magnifies. The Code of Practice governing the broiler chicken industry stipulates that the birds cannot be moved if the temperature exceeds 40°C. Transport vehicles are also expected to offer wind and weather protection. The birds are picked up by the legs and put into specially designed plastic crates.

Most trucks arrive at the processing plant sometime after midnight, or at times determined in the planning ordinance laid down by the local Council. The plant may be sited in an industrial park or across the road from residential homes. Depending on the temperature and the wind, one may be struck by an unpleasant smell and the noise of the trucks coming and going.

What is strikingly different between the farm and the processor plant is the hustle and bustle. The farm is quiet, whereas the processing plant has people all around. The vast majority are in uniforms of white coats, galoshes, ear plugs and hairnets or hardhats. There is a sense of urgency. From killing to chilling ideally takes thirty minutes in order to prevent blotchy flesh and other undesirable features of meat quality. There are also vast quantities of 'fresh' product to process in one shift to meet tomorrow's order for supermarkets. One processing plant may be expected to supply one supermarket chain with three and a half tonnes of breast fillet, three tonnes of thighs, the same amount of wings and four tonnes of drumsticks, five days a week.

The pressure is even greater in summer. This is when demand for chicken skyrockets and 'you've got a live animal, it's a hot day and all the birds are dying on you or a machine breaks down and you know everyone will be wanting chicken take aways'.[6]

In the back landing of the processing plant the birds are unloaded by several men who have a difficult job keeping the birds inside the crates. Handling the smaller birds are four male hangers wearing masks, gloves and plastic aprons. They lift the flapping birds upside down onto the shackles, attached to a head high conveyor belt. This team is capable of handling 5700 birds in an hour, while feathers and dust fly, and faeces and urine spray around them. Another team of three handlers is assigned to the large birds and they deal with sixty-five birds per minute or 3900 an hour. The two teams rotate throughout the day.

The hangers handle the birds calmly and with a certain easy style borne of practice. In spite of the introduction of dust extractors and the like to improve working conditions, the nature of this task has not changed in thirty years. Some plants are introducing blue lighting in this section to further quieten down the birds.

Hundreds of flapping birds move overhead into a stainless steel chamber where their heads and necks pass through electrified water. The now stunned birds move quickly through another machine whose mechanical knives cut the jugular vein. A person with knife in hand stands nearby to kill any bird which escapes the slit in the neck. The bleeding after stunning is contentious in terms of the final product. The processors believe that incomplete bleeding produces meat with a dark colour that is tougher. One poultry industry body submitted to the Senate Committee on Animal Welfare that stunned birds bleed well, dead birds do not, and it 'was important in the final product that the birds, in a stunned state, bleed out before they die' (SSCAW 1990, p.156). In a counter argument, the Federation of Animal Societies urged that the voltage of the stunner should be raised to a level where all the birds are killed.[7]

After stunning and neck slitting, the birds travel above steel drains as they bleed out. The birds go through a scalding process to facilitate feather removal, which is completed by a defeatherer, or a system of rubber flails. Before long their heads are efficiently removed by another machine. The observer quickly realises the critical importance of having birds of equal size so that the machines can be set to do their job. Even though this plant has a small bird and large bird line, a man hovers to make adjustments to the decapitation machine to ensure it removes the correct amount of head and neck.

Owing to the high temperatures at this stage, the ensuing water bath may contain chloride to kill bacteria that flourish under heat. Making the meat hygienic in this manner has another benefit: white chicken is more desirable from a consumer perspective, although the focus group participants did not know about this cause-effect relationship. It is possible to obtain 'unbleached' chicken meat from a few smaller processors who do not scald the birds — the meat of the resulting chickens is pinker than otherwise, but because the defeathering stage is more labour intensive the chickens are more costly.[8] A considerable amount of water is used to wash away the feathers: the latter being recirculated in animal feed and litter.

At this point one is well and truly conscious of three things: constant noise of the machines, bloody water everywhere — this is not to suggest that floor hygiene is compromised — and lots and lots of birds circulating overhead with very few people. The rather strange sight of a line of feet moving overhead was a relief to me because the live animal of two minutes ago is clearly dead and quickly resembling chicken meat products. I am pleased to note that my emotions are not unique. In a description, from the 1980s, of a meat processing plant in New Zealand, the researchers noted:

> [t]o tour a modern export meat-freezing works is an enlightening, though somewhat gruesome, experience. The first thing that someone who hasn't been there before notices are the sights and smells of the product: bright spouting blood, stripped shining flesh, half severed heads dangling grotesquely from swinging carcasses, steaming entrails and internal organs tossed casually into shining steel trays. There's a quality of hygiene nowadays that wasn't always there ... But the noises are of the factory [not the clinic]: pounding motors driving the clanking production lines which move the product relentlessly on (Inkson & Cammock 1988, p. 69)

While the chicken processor is dealing with much smaller animals there are marked similarities with the preceding description as the birds pass onto one of two evisceration lines, again depending on their size. This is where the noise is at its peak, with machines everywhere conveying carcasses with varying amounts of gut hanging out. Standing on slightly raised platforms is the lone woman or pair of women deftly removing organs: the livers are being dropped into plastic containers, the heart, lung and other bits go into further containers, destined for pet food. On this day, one woman was sandwiched between three moving lines manually removing feathers left by the defeatherer: some days the machines are not as efficient at their job as other days. In fact, in the evisceration room machines appear to have let management down: this section contained thirty-four women ten years ago and the new machines were meant to cut that to four, however fifteen are required to get the tasks done. So while there is a machine to remove the remaining viscera, it misses a lot and one worker's job was to plunge her fists into each bird and pull out what remained. I was told that this is very hard on the arms and shoulders and for this reason workers in this section are rotated every half hour.[9]

The orders placed with the plant will determine to some extent the speed of the line and on the day of my visit the pace was fast. Again the fit between bird size and machines is of importance to the workers, as is the maintenance of machines. Daily maintenance, rather than breakdown maintenance, was considered by process line workers to be vital to efficiency and work satisfaction on these lines. A poorly functioning machine is obviously worse than no machine at all and a couple of workers reminisced about the old days when there were more workers, less speed associated with the job and fewer machines. One thing is for sure; talking to one another is almost impossible in this section. It's just a matter of silently exercising repetitive eye-hand coordination over long periods of time, punctuated by a ten minute morning and afternoon break, a lunch period of thirty minutes and, on this site, a seven minute 'smoko' allowance.

Break periods at this site, which has the best canteen and bathrooms in the state, appeared to be very convivial occasions with much sitting in groups and guffaws of laughter. The breaks provide opportunities to approach the shop stewards about matters of concern like, 'Why can't we have two pairs of gloves in a shift? Mine get wet, then my fingers go numb and I can't do the job', or, 'I want to resign in two weeks, what does this mean in terms of my entitlements'. Sometimes small groups approach and ask if something can be done about a particular machine.

Back to work. Further large volumes of water are used to spray the birds to remove any stubborn gunk and to minimise bacterial deposits on the skin. Chilling follows promptly and large amounts of ice sit ready to be shovelled onto conveyors for the trip to the next machine. Most birds are rehung for the boning, bagging and marinade room, where the noise level is low enough to have a radio belting out. Once the orders for uncooked chickens have been met, some special requests are filled, for example, the Chinese New Year orders for chickens with head but no feet. Some processors accuse their rivals of placing the better looking birds in bags destined to be sold as free-range, even though they are not. A sizeable proportion of chickens will be cut up and marinaded. The birds pass along conveyor belts to be sorted and graded and those to be boned are placed in rapid succession onto small plastic cones where they receive highly skilled cuts. Breast meat is removed, legs are separated from torsos, thigh meat from bones and so forth. To watch some of the boners is sheer pleasure because they get into a rhythm of cuts, and make cutting raw meat seem effortless. Newer staff seem to be more ungainly and tentative: they complain most about blunt knives.[10] Once again my mixed emotions for what I was witnessing — admiration and relief that this was not my job — were supported by the New Zealand researchers quoted earlier. They commented '[a]midst [the machines] are the workers, busily pulling, cutting, trimming, washing, grading; fingers working, knives flashing, over and over again, the same precise, practised, drilled sequence: a job cycle over in seconds, active but mindless and automatic' (Inkson & Cammock 1988, p. 69).

No process workers like the big birds: they are heavier to hang, the machines do not handle them as well, especially in the evisceration area, and they 'require deeper, longer and more forceful cuts' (OHSA Manual n.d., p. 24). This is not helped when machines designed to hold 2.6 kilogram birds must handle birds up to 3.1 kilograms. Nevertheless, those on the boning lines are pleased today because the line speed has dropped from nineteen to eighteen birds per minute and the workers rotate every half-hour. The leading hand in this section expresses satisfaction with the machinery introduced over the last six years: 'It's made the job easier'.

Back to the processing plant, and a smell of chicken noodle soup wafts up from the marinade injection machine. Here, 'the eleven herbs and spices' are being pin-pricked under the skin for a batch destined for KFC. A line of nugget-size pieces comes down a short conveyor belt covered in dark sauce marinade. About twelve workers are engaged in putting pieces onto Styrofoam trays, weighing them, covering with plastic, labelling, boxing and organising the orders for supermarket chains, fast food outlets and others. Again they work hurriedly, moving between machines, with no time for chitchat. Men are coming and going with trolleys taking the final product to the cold store. There the dress is woolly beanies and padded jackets.

Everyone knows that by knock-off time at 3.06 pm, the regular end of the day's shift, their seven hour thirty-six minute shift of 260 workers will have killed and processed 58 000 birds.[11] This gives many a sense of satisfaction and a number will be preparing chicken for the evening meal, bought either from the on-site store or from the supermarket on the way home.

Trucks leave the processor from 2 am onwards, to deliver whole and portioned chickens to supermarkets, fast food outlet warehouses and specialist poultry shops. A special contractor will arrive to pick up the carcases and innards and take these to a rending plant, although by-product is rendered on some of the large sites. This procedure causes some worries and, due to fear of cross-contamination of bacteria, the contractor will be detoxed at the gates of the processing plant, in case he has been on an 'unclean' farm removing its dead birds.

Between pick up and death, the birds are alive for between two and twelve hours, and the process of turning raw meat into 'fresh' chicken meat products takes less than an hour. This, as they say in the industry, is working with 'real time'.

And what about real money? I did not hear complaints from any processors about the hourly rate of pay, but they were keen to reduce the size of their labour forces. The introduction of machinery to replace workers seems an across-the-board priority, in order to gain further efficiencies. As one processor put it, 'We would prefer to spend one to one and a half million dollars, the cost of setting up another company farm, on another boning machine. In that way we'd dispense with another thirty workers. That represents a better return on investment'.

Previous accounts of labour processes have generally attributed the diminishing power of labour to the introduction of chain processing systems (Inkson & Cammock 1988; Mathews 1989). In the meat industry, Mathews notes the successive displacement of skilled butchers by machines and the hiring of unskilled labour. He also describes the inefficiency of the chain assembly lines since their introduction and, as the present fieldwork shows, they are still less efficient than processors would hope. It seems that the assembly lines introduced by Ford were about control of labour as much as efficient production: defining jobs in narrow categories, defining jobs by the machine used rather than through the desired output and deskilling in order to cheapen labour (Mathews 1989, p. 123). However, organised labour in meat and poultry processing is alive and well. The workers strength *en masse* is created by two quite different features. The first is bipartite: a chain system by its very nature forces an interdependence among workers; while its set up and running costs limit its existence to only a few sites, thereby creating a concentration of workers. Under these circumstances, worker organisation flourishes. The second is bargaining strength: the perishable nature of the product means management relies on those workers to ensure the smooth running of the plant's operation. It is in this context that it has been argued 'the practices of successful modern management in the [meat processing] industry are therefore likely not to be those of coercion ... but those of "responsible autonomy of workers"' (Inkson & Cammock 1988, p. 73).

Besides their desire for machine labour to replace human labour, enterprise agreements are increasingly favoured by employers to replace union awards. One chicken meat processor explained the situation in post-Fordist, or flexible labour, terms:

> In the value-added plant we have about sixty employees who we pay on performance — which is miles above the award, it's twenty-five per cent for some of them, but it varies on the output of the worker. Since we've done that our productivity is up about ten per cent, our quality is fifty per cent better and our return from customers is down to .01%. We are rewarding some way above the norm. We need to provide a sense of belonging and second a sense of self-worth. No award can give you any of these things. An award takes away the individuality.

This is the context in which union stewards work: knowing management is constantly looking to cut costs by hiring the least number of workers without compromising product quality. When permanent staff are away, casual staff from specialist agencies come in. They are not only paid a substantial amount more per hour, but they have no commitment to the firm or the site. This practice, it seems, is the beginning of creating numerical flexibility in poultry processing.

The priority for a union delegate in an era of deregulated labour markets is not to lose the hard-won gains for workers and to constantly educate the members as to the pressures on the industry. Changes within the industry are making the organising role harder. Some union officials express concern for the plants with large concentrations of non-English speaking workers: an increasing feature of the processing side of the industry. Thirty years ago, it was predominantly an Anglo-Australian dominated workforce. Now, depending on the plant location, one can find an almost exclusively Asian or Macedonian labour force. The Anglo-Australian union officials point out that new immigrants' concern for employment conditions are less than English speaking workers, with greater risk of getting less take home pay and accepting worse working conditions.

PRODUCTION PRACTICES OUTSIDE THE COMPLEX

Glenloth Poultry in rural Victoria grow free-range chickens in a kit shed worth about $8000, unlike the sheds described earlier which cost hundreds of thousands of dollars. In this particular shed 2000 birds coexist at any one time without air conditioning or special ventilation systems. For the first four weeks the growing chickens do not leave the shed but are brooded under lights. For the remaining four weeks they are let out into an enclosed field during daylight hours: only for half of their lives do they free-range. Every effort is made to use local grain, but the cost of the locally sourced feed is determined by prices and conditions in Queensland, where the bulk of the eastern states corn is grown. Free-range birds generally take longer to reach the two kilogram weight and are often killed at about 1.5 kilograms. Medication is rarely required because this is not an intensive growing situation, acknowledged as the harbinger of infection.

The killing process is vastly different in this tiny operation which processes only 300 birds a week. Two people herd the birds and gently put them into crates: 'unlike the ten at a time operation, because otherwise we bruise them and we can't afford to lose one'. The crates travel on the back of a trailer across the field to the processing shed. The birds are stunned in the same manner as in the intensive system, but because of the uneven sizes of the chickens they are killed by cutting the necks by hand. They then pass through the defeatherer. 'We don't have economies of scale, so we try to operate on top quality produce — that means strict reject rates. If people are paying $10 to $11, as compared to $5, for the chicken they don't want a bruised one'.

The relative absence of machinery (no mechanical knives are used), the personalised handling of the commodity by the owner-operators and an obsession with quality are hallmarks of craft production: a type of production aped within post-Fordist regimes of accumulation.

At Glenloth, it is Chris, one of the women business partners who kills the birds: a certain irony because whilst living in the city she was vice-president of the RSPCA (Wildlife). These particular farmer-processors believe that women are better with animals: 'They're more patient, [have] better powers of observation and detail and the birds react differently to women'. Chris told of how she joined the business because she liked animals and was interested in breeding pheasants. 'I couldn't kill anything. But one day a bird was going to go through the scalder alive if I didn't rush over and do something about it. After that ... you can't decide to make your living out of birds, but not kill them'. As these owner-operators put it: 'concern for the birds is good economics. If we don't look after the birds, then we get a poor result and go broke. If the birds are neglected, then it hits you a couple of weeks down the track'. Thus in a curious way these niche market producers care more about animal welfare issues than the consumers whom I interviewed. What was made clear is that the minuscule market for free-range products would evaporate if the appearance of Glenloth birds, which retail for at least twice the price of other chickens, is contravened. What remains unclear from the research is whether the free-range niche market is a new market or whether it is an extension of a market containing consumers familiar with the backyard roast chook. The absence of free-range chicken meat consumption in my focus groups, most of whom were familiar with the backyard chicken shed, suggests that it is a new market, possibly consisting of consumers nostalgic for a past that they did not experience. The focus groups, however, suggest that health issues would be a potent reason for them to shift from the mass produced to the flexibly produced product, if the price was comparable.

PRODUCT DESIGN

Processors generally attribute their success to their early awareness that to be profitable they have had to 'even out', or use the entire product. Thirty years ago that meant finding uses for the breast meat that was left over after the drumsticks had been sent to KFC and other chicken cuts had been despatched to hotels for chicken Maryland and 'basket suppers'. The ever-changing evening-out challenge persists into the 21st century: supermarkets continue to demand increasing amounts of deboned meat to sell both as whole or half fillets or to value-add with condiments leaving processors with stockpiles of wings. 'I wish they would produce a one-winged bird' is a mantra among processors.

An examination of the evolution in chicken meat product

development over the last thirty years gives valuable insights into the importance of product differentiation strategies and the shifting balance of power within the complex. In the 1960s product development was a pretty basic affair, as described by one processor's manager:

> I'd get an idea and know where to find the director — in the canteen at lunch time. He'd scribble a few notes on the back of a piece of paper and say go for it. Now its detailed business plans, showing that we can pass the hurdle rate of four years [that is repay the investment in that time]. Also, go back ten years and customers accepted shortfalls, or inconsistent sized products. Not now; the customer wants the same bloody product day in, day out.

In the late 1960s, Steggles was the first major company to build a dedicated value-adding factory in Australia enabling it to supply more than frozen and whole birds. Some staff claim that the company really started to make money as a result of the Chicken Roll: an idea that Bruce Steggles conceived on one of his overseas trips. The story from a food technology manager who worked on the product follows:

> I was working for a poultry processor in Scotland twenty-five years ago, but was unhappy about the way that things were going in the United Kingdom, so I wrote to similar firms here. Steggles invited me out here to turn the Chicken Roll into a production line item. The need [for the product] arose because the amount of product being sold in portions, mainly wings and legs, left a surplus of breast and thigh meat. There was no research conducted on any products back then. The original trials were conducted by producing the product then sending it out to our sales force and hoping that it sold! The concept for the Chicken Roll was adapted from the red meat industry.[12] Although the production methods were a secret, the hardest part and the most secretive area was in the deboning of the chicken on a moving line. There was no expertise back then ... all the staff had to be trained from the beginning.

The first few production batches did not sell until a food technologist took it upon himself to approach a smallgoods red meat producer for marketing assistance. The first Chicken Roll sold in Australia was produced in the well-recognised DONS Smallgoods casing. This manager was proud of his contribution to this particular food and of the role the Chicken Roll has played in the company's success, but noted with some sadness that it would not be repeated:

> It was originally a good product, put on the side of the plate besides salads in hotels and restaurants. It was a good product, even though a by-product because it was in natural proportion. Only that amount of skin that was attached to the breast and the thigh was used. Now it is perceived to be like devon [fritz], and is part of the gut filling market. It is produced by many processors, and is probably extended enormously by extra fat, skin and other ingredients. Skin is only forty cents a kilo while breast is three dollars per kilo — skin adds flavour.

The greater complexity and expanded number of players involved in today's product development process is illustrated by another of Steggles' products, designed for the food service sector. In mid-1993, the Steggles Sales and Marketing Manager returned from a visit to the United Kingdom where he was impressed by a number of poultry and meat products encased in puff pastry. He prepared a concept statement and convened a meeting with the relevant manufacturing centres within Goodman Fielder to discuss the product's feasibility. By the time the product was ready for a test market thirty-five weeks after this initial meeting, the development process had incorporated the Steggles Product Development Committee to approve the concept, food technologists from the Steggles Foods and Ingredients Suppliers division to submit sauce flavour varieties, the marketing division to evaluate these varieties, and the sales development manager to organise production trials with the pastry supplier and produce concept samples.

According to Simon, a food technologist associated with the product, 'no special technology was required, but the product was unique for Steggles in that technology associated with chicken was merged with technology related to pastry manufacture to produce a novel concept'. Steggles named its new product the Pocket Rocket, to be promoted in school canteens on the basis of novelty, lower fat content than a beef pie and taste appeal.

Like the Chicken Roll, pastry encrusted chicken was a concept borrowed from overseas markets and was not the outcome of any local market research. The tasks and people involved in this particular 'me-too' product are typical of product development in the 1990s. While the Chicken Roll and Pocket Rocket came from a large processor, the supermarkets argue that the smaller firms are today's innovators. I was continually told that poultry is an economies of scale business, incorporating two types of market leader: the suppliers of cooked fast food and fresh chickens for supermarkets, and the fresh, value-added product suppliers. Inghams and Steggles, with their nationwide operations, provide for the first customers while some medium sized processors lead the way in the second market. A co-owner of Marven Poultry provided a rationale for how this state of affairs has evolved and in so doing also revealed that consumers have had minimal influence on product development. She began by explaining that:

> [a]ll I've got to sell is that we're innovative and that we're the best. I've got to back that up with the product. The big processors are the market leaders in the cooked fast food value-added market, because that's very capital intensive and we can't afford that. And the fresh value-added is very labour intensive — they can't compete with us because they are further away from their direct floor labour.

She explained how survival by the second tier processors could not be ignored as a factor driving product development:

> Our need to grow is enormous — to expand and to keep financing our expansion rather than to get the volume of finance from a financier means that I have to have a minimum five per cent growth rate per annum at a set return, so I can invest my 2 and 2.5 million dollars each year in new infrastructure and keep at the very front of technology. I don't have to have Coles telling me what new products they want, because to continue to be at the forefront, that's the growth I need. You have to be innovative.

This particular product innovator brings ideas back from Germany, her place of birth, and produces a smallgoods range aimed at migrants from Northern Europe. To do this she employs specialist food technologists and has a dedicated factory for highly processed products. Her company also has entered into arrangements with other small processors to share the costs of a smoke-house and additional processing equipment to achieve economies of scale for niche marketed products. She acknowledges that such products are not considered staples and that trading in novelties and niche lines is more risky than trading in the mass market. But being able to deliver variety is critical to the success of the poultry industry in her opinion because 'what the red meat industry did was to put a big bit of red meat on the plate, and that big bit is still sitting there. It's only now being cut up — they haven't matured. It's where we were twelve years ago'.[13] For this process owner, product variety is what distinguishes the red and white meat industries as much as the differential health claims that some experts allege are responsible for chicken's esteem.

Product development is in the process of being transformed and it is retailers who are telling the processors which products they want. While Coles or Woolworths may not dictate the whole product range it appears that they play a far larger role than consumers. In comparing trends over the last twenty-five years one food technologist considered the best products to be those that are the easiest to make because they involve the fewest steps. Often the cheapest products in the supermarkets, like the Pocket Rocket, are the most complex and require more staff on behalf of the processor, 'but we do them because this is what the supermarkets order. The consumer in the supermarket has no choice — they get what's in the supermarket. We produce things that cost us more because the supermarket want them and they are our bread and butter'. This view of the relative power of the consumer and the supermarket was reinforced by the manager of the largest processor in Victoria, who said that consumers 'buy what is available — they're a secondary influence on what is produced. The retail customers are the primary influence'.

Indeed the only occasion on which I heard processors acknowledge the influence of the consumer on product development related to the need to consider products which delivered convenience and health. One said:

> [e]very product follows social changes. Everyone wants a leaner, healthier product which is convenient to eat — it's about minimising work for the housewife. We give housewives products which make her think she is doing something. [For example,] there has been a twenty per cent increase in filleted products and a huge demand for heat and serve dishes.

Paradoxically this is the same processor who bemoaned the fact that 'the consumer wants the same bloody thing, day in day out'. Here we glimpse tensions between a Fordist consumer market and a retail customer market that demands novelty. As one supermarket buyer commented, 'If there is not one new product a month then it has been a bad month'. Certainly no one I encountered questioned how convenience, health and novelty can coexist in industrialised food products. Herein lies the basis for the need for clever handling of the commodity context.

CONCLUSION

When assessing the balance of power between primary and secondary producers, producers and retailers, and producers and consumers, numerous ambiguities arise. Clearly, defining the chicken meat complex in terms of regulation theory — especially in Fordist/post-Fordist terms — is problematic.

Chicken farmers are engaged in an arrangement that is historically unprecedented. Those who perform the labour of growing chickens are simultaneously self-employed contract labourers and 'landed labour' by virtue of their ownership and investments in their own properties and expensive machinery (Davis 1980). While hardly craft producers in the sense of producing small, specialised and differential production runs, the chicken meat farmers have to be multiskilled and extremely knowledgeable about their occupation. Moreover, they can exert some flexibility over their labour routines without necessarily jeopardising their livelihoods. Their labour is self-regulated rather than machine regulated, even though much of the labour occurs in factory-like conditions. This chapter indicates that it is premature to classify all of poultry production as Just In Time because this would overlook 'the real time' characteristic of an industry that is based on looking after tens of thousands of live creatures subject to the vagaries of genetics, weather and disease. The personal attention that is paid to birds, particularly ailing birds, adds an element of services work missing in other more durable agrifood commodities and one that led me to a profound respect for those involved in commercial chicken production.

It is far easier to characterise the activities of the secondary producers or processors. Their processing plants are exemplars of Fordist production methods delivering mass products day-in day-out regardless of the season. This fact should come as little surprise when one notes the assembly line that gave its name to the Fordist regime of production came originally from the meat industry. Apparently Henry Ford obtained the idea of chain methods of automobile assembly from observing the chain methods of dis-assembly used in meat processing works around Chicago in the 1890s (Mathews 1989, p. 25).

If Fordism reigns in the factory, a more curious arrangement operates between processors, farmers and retailers. While the standardisation and homogenisation of the chickens are clearly Fordist characteristics, the use of numerous batch producers (the farmers) who are locked into Just In Time delivery schedules with the processors resembles post-Fordism. The Just In Time practice is particularly pronounced in the way that processors take birds off farms on a variable basis and in line with daily demands of their large customers, supermarkets and fast food chains. Post-Fordism, it seems, is present in the supply chain via Just In Time procurement from suppliers.

What of product development? The extraordinary range of chicken products gives the appearance of a differentiated product range and, in the eyes of producers and consumers, product differentiation distinguishes the chicken meat industry from the red meat industry. However, in the Australian context at least, Kim and Curry (1993) were correct to note that what often is called post-Fordism is the mass production of variety. The majority of what appears niche product is the same mass produced cuts of chicken, sauced and coated and mixed with other ingredients in myriad ways.

A lack of real alternatives in chicken meat products, such as affordable free-range chicken and 'healthy' chicken, leaves one with an impression that consumers have little influence over the production of chicken meat products. Or maybe they do have influence: by buying essentially the 'same bloody product every day' they signal appreciation of the Fordist regime. Consumer indifference and/or ignorance about production seemed, from the focus groups, to generate misplaced fears about what they were eating while stifling demand for products that would better meet their needs. In contrast, consumers appear to have had a bigger impact on the red meat industry. By not purchasing the amounts of red meat that they used to and by demanding the dis-assembly of red meat into smaller pieces and its re-assembly into more convenient, healthy products they are demanding that red meat become more like chicken!

Finally, this chapter bolsters a proposition that emerged from the last chapter: producers and consumers appear to inhabit distinctive worlds when it comes to the table chicken and such is their separation that there is little tension between them. Instead, the more significant tensions are between processors and contract growers and between processors and supermarkets. The second of these tensions forms a critical part of the next chapter.

6

'HERE A CHOOK, THERE A CHOOK, EVERYWHERE A CHOOK CHOOK'[1]

Chicken, it seems, was everywhere in the late 1990s and nowhere more so than in shopping malls. Take the Knox Shopping Town in outer Melbourne, where chicken could be bought from one of the following outlets: Lenards' specialist poulterer; three butchers (one of which had more chicken than red meat); the gourmet sausage stall; and the Bi-Lo supermarket. At Coles chicken was available from the delicatessen (hereafter referred to as 'deli') section, specialist poultry section, fresh meat section, dairy cabinet for frozen birds or one of the many cabinets containing pre-prepared meals such as Lean Cuisine. If all that required too much effort shoppers could walk over to the Red Rooster outlet and buy a cooked quarter chicken and chips or have a chicken sandwich in any one of a dozen coffee shops dotted along the three levels of the complex. Except for free-range produce, the choice of chicken products appeared overwhelming. Not so daunting was the available selection of fish, which was offered at the two supermarkets and only one specialist fish shop. There was certainly no beef or lamb equivalent to the specialist poulterer.

Herein lies the challenge of describing chicken meat distribution in a complex capitalist marketplace.[2] The sections that follow are organised according to the commodity analysis framework outlined in Chapter 3. The material concentrates on retailers rather than wholesalers, because, as seen in the previous chapter, the latter are irrelevant to chicken meat. It would have been logical to detail the operation of fast food chains in this chapter under the 'Food Services Sector' heading: after supermarkets, fast food chains and independent take away stores sell more chicken than any other type of retailer. Supermarkets receive most attention here because, as became clear in my interviews with primary and secondary producers, they appear more responsible for chicken meat's popularity than any other group. The strategic positioning between supermarkets and specialist poulterers is used in this chapter to examine the effect of retail accumulation strategies on the circulation of chicken meat. These strategies are increasingly based on shaping the commodity context through product differentiated retail systems. Such systems require the restructuring of supply chains between producers and retailers and necessitate the moving of retail capital into the circuit of production. These concepts, set out in Chapter 3, are explained here through vignettes of the people and processes involved in selling chicken. The labour process of two categories of supermarket employee — the national poultry buyer and delicatessen manager — is outlined. The latter are mainly female employees who are pivotal to the distribution and exchange of chicken at the point where production and consumption intersect. Theirs is a world of pressure based on supplying other pressured food providers with what has become a family staple. I conclude with some remarks on the distinctive lack of regulatory codes in the sphere of distribution.

FOOD DISTRIBUTION IN AUSTRALIA

Australia's agrifood industry — which includes agriculture and fisheries, and the processed food and beverages industry — was worth $60 billion in total retail sales and exports in 1997. Food production accounts for around six per cent of GDP, approximately seven per cent of total employment (estimated to be 1.1 million people) and around one-quarter of all goods and services exported from Australia (Australian Food Council Fact Sheet n.d.). Food distribution takes place via two major channels: supermarkets and the food services sector, which is made up of an institutional sector of hospitals, nursing homes, educational institutions and prisons; and a commercial sector of restaurants/cafes, hotels/motels, clubs, fast food chains/independents and catering contractors. If one includes the number of people involved in food wholesaling and retailing, the combined business of food production, processing and distribution constitutes Australia's single biggest industry sector, employing fourteen per cent of Australians. The Australian Chicken Meat Federation estimates that nearly 60 000 people are employed to retail chicken meat (Fairbrother 2001).

FOOD RETAILER PRACTICES AND ORGANISATION

This section focuses on two types of retailer that play a significant role in relation to chicken meat: supermarket chains and small specialist poulterers. While processors credit supermarkets with maintaining chicken's low retail price, they express gratitude to specialist poulterers for their chicken meat innovation. The material in this and the following chapter undermine arguments about an abundance of retail forms. Alternative retail markets based on different retail practices, like the choice of alternatively produced chicken meat, are somewhat illusory.

SUPERMARKETS

The 1996 Tenth Australian Poultry and Feed Convention was a convivial affair attended by over 600 of the nation's egg and chicken meat producers and feed growers. While the opening address by the acting Victorian Minister for Agriculture was greeted with a world weary cynicism that one imagines farmers reserve for politicians, it was both the speech by the Franklins Director of International Fresh Food Development and the address by Woolworths Manager of Retail Operations that gained the delegates' respectful attention.

Both executives portrayed Australia's future as 'healthy, strong and vigorous', with their corporations playing an important leadership role. They congratulated the primary producers on the quality of their product. However, both told the delegates 'to get real': business growth was becoming contingent on producers being more customer oriented. Here they were referring to retailers rather than household consumers. An improved customer orientation, they opined, entails industry restructuring to meet the forthcoming challenges of competition that will emanate from the arrival of imported products.

The only commodity outsider led the other popular session. 'The persecuted pig' was the title of the paper given by the President of the Pork Council of Australia (PCA) and it portrayed supermarkets in a different light. The audience heard that the supermarkets' procurement of Canadian pork, coupled with the 1995 drought, forced twenty per cent of Australian pork producers out of the industry in a single year. Despite a consumer backlash forcing the Woolworths chain to publicly announce that it would no longer accept Canadian pork products, the speaker disclosed that all major supermarket chains maintained a policy of using imported product to counter domestic undersupply. The PCA's assessment of its future was upbeat though, based on plans to increase domestic consumer's taste for pig meat and becoming exporters to Asia. This snapshot encapsulates the importance of the supermarket chains to Australia's agrifood commodity sectors. Whilst farmers and food manufacturers remain critical to the commodity complex, those in charge of the distribution channels and of government regulatory policy are leading the current restructuring in those complexes. Increasingly, as we will see below, supermarkets are exercising the balance of power in chicken meat distribution channels, although their power over food distribution is not going unchallenged by farmers or by smaller food retailers.

In late 1998 the NSW Farmers' Association foreshadowed an investigation into the buying power of the Woolworths, Coles and Franklins supermarket conglomerates, because of their monoploy of more than sixty per cent of the fresh produce market (NSW Farmers News Release 1998). As part of their campaign 'Enough is Enough', the Association produced a breakdown between independent stores and the major chains, showing the rapid replacement of the former as a point of supply for groceries. In 1975 independent food retailers accounted for sixty per cent of grocery trade but by 1997 their share was down to twenty-two per cent and falling. This trend to oligopoly control by the three supermarket chains sparked yet another far-reaching investigation into food retailing. This time, the powerful Joint Select Committee on the Retailing Sector (JSC on the RS) was asked by the Commonwealth Parliament to investigate the degree of retail industry concentration and the ability of small independent retailers to compete fairly. The inquiry originated from pressure by the National Association of Retail Grocers of Australia (NARGA), who reasoned that the demise of hundreds of small grocers, butchers and green grocers, and the process of globalisation which has 'seen ... primary producers having to compete for markets not only against their fellow Australians, but with others in the same business around the world' (JSC on the RS 1999, p.1) warranted scrutiny. It is noteworthy that consumers were not deemed to be a force in the establishment of the inquiry; rather the JSC viewed them as the major beneficiaries of recent retail developments.[3]

The Committee Report released in 1999 reveals the significance of food retailing within retailing generally, with supermarkets and grocery stores, take away food and other food stores accounting for forty per cent of total retail turnover in 1998. Furthermore it provided data that confirmed the misgivings of the NSW Farmers' Association. The Committee produced figures which showed that the three largest supermarkets traded in over eighty percent of the grocery market in that same year. Table 6.1 reproduces the state of play.

Table 6.1
Australian food retail industry concentration

	Woolworths	Coles/Bi-Lo	Franklins	Total
NSW	36.4	23.4	24.2	84.0
Vic	36.6	33.8	8.7	79.1
Qld	38.6	32.2	16.4	87.2
SA	29.9	38.0	7.0	74.9
WA	27.1	33.4	n/a	60.5
Tas	73.1	26.9	n/a	100.0
National	35.9	30.3	14.2	80.4

SOURCE Joint Select Committee on the Retailing Sector 1999, p. 39.

Eighteen months after the Committee reported, Dairy International, the Hong Kong based owner of Franklins, put all of its stores on the market. Woolworths, already the largest food retailer, bought a significant number of the chain's 270 outlets. Coles and the independent sector purchased the remainder. Thus Table 6.1 underestimates the extent of industry concentration, which in 1998 far exceeded the situation in the United States and Britain. The figures in Table 6.2 show market share by the leading supermarket chains in 1992 and 1998 for Australia and the United States (Burch & Goss 1999). Other figures for Britain (Hughes 1996, p. 97) confirm that the level of concentration in Australia outstrip by two to three times the two nations on whom the Australian operators have modelled themselves.

Table 6.2
Concentration in the retail food sector in Australia and the United States in 1992 and 1998.

AUSTRALIA	1992	1998	UNITED STATES	1992	1998
Woolworths	29.3	35.2	Kroger	5.8	6.4
Coles	21.8	28.5	American Stores	5.0	8.1
Franklins	15.1	14.7	Safeway	4.0	6.5
Total	66.2	78.4	Total	14.8	21.0

SOURCE Burch & Goss 1999, p. 337.

Woolworths became the nation's biggest food retailer in 1985 when it bought Safeway stores from their American parent. Coles may not be the biggest food retailer but it is Australia's largest retailer overall: in the mid-1990s the company commandeered twenty cents in every retail dollar spent, making it 'the most powerful retailer in the world on a per capita basis' (Gawenda 1996, p. 28). Among its food interests, Coles also owns Red Rooster, the nation's second biggest chicken take away chain.

However, it is the longstanding and enthusiastic supply of table chicken that leads the producers I spoke with to credit Woolworths and Coles with assisting the processors to make chicken Australia's second most preferred meat in such a brief time. The supermarket strategy of loss-lead marketing, or the ability to identify products that will *pull* in consumers and then *push* them further into the supermarket, is well known. Both the bacon rasher (which includes the fillet of the pig) and the chicken fillet join a list of about eight other items, including margarine and Coca-Cola, as the pre-eminent loss leaders for supermarkets.[4] These items retail for close to, or below, production costs to increase both consumer spending, and profits made by producers and retailers on related and other products.[5] One processor's national marketing manager told me that 'supermarkets live and die by the fillet'.

Conversations with supermarket executives confirm that over the years, chicken meat has been used by supermarkets to attract customers. That chicken is regarded as part of the Australian supermarket success story was supported by the front cover headline of a Franklins brochure, '*searching ... for a challenging career*?' It depicts a rooster driving a rocket and the image is accompanied by the statement, 'The rocket — bestraddled by a chicken — illustrates that we have the Vision, the Mission and the team Values that will underpin our strategies to develop Franklins towards the year 2000'.[6] Prior to its sale, Franklins had clearly been competing with its older, larger rivals via 'the fresh chicken concept'. Chicken is the premier symbol of fresh food for middle-aged and older Australians due to its transformation in the 1960s from a frozen into a chilled product that was perceived to be fresh and natural. Some would argue of course that frozen and minimally processed foods, such as eviscerated chickens, are closer to nature than the sauced, marinaded and otherwise value-added heat-and-serve meals that are currently sold from the Deli-fresh sections of supermarkets.

So it is interesting to learn of the genesis of 'fresh' food. It seems that consumer demand for fresh produce had little to do with the adoption of the fresh concept in Australia. One Woolworths CEO noted that his company turned in 1983 to 'fresh' in the quest to distinguish Woolies from its rivals, just as supermarket chains were doing in the United States (Shoebridge 1994, p. 43). As a marketing strategy the fresh concept has been very successful: in 1994 fresh food accounted for forty per cent of sales in Woolworths' supermarkets, up from twenty per cent in 1987 (Shoebridge 1994, p. 40).[7] And Woolworths' expansion at the expense of Coles appears attributable to it becoming the market leader in fresh foods, because it provided a head start on a trend that gathered momentum in the mid-1990s. As *Supermarket* magazine remarked, '[t]he 1996 fight for the grocery dollar will be as vicious as 1995. The focus now is on each player beefing-up their own offering with the word FRESH paramount in any store design or marketing strategy' (Flanagan 1995, p. 17).

More recently, the Australian Supermarket Institute (ASI) provided insights to the supermarket chains on how they could compete in the fresh food market against butchers, greengrocers, bakers and delis (ASI/AC Nielsen 1998). Based on consumer research and 'anthropological data' we learn that female shoppers prefer to purchase meat, or at least red meat, from a butcher because of a factor that is called 'familiarity': 'the butcher [is] a paternal figure who provides shoppers with a sense of trust and expertise' (ASI/AC Nielsen 1998, p. 52). And while a large number of shoppers — the 'social-shopping' and 'resentful-shopping' segments — rebel against familiarity and buy their meat in supermarkets, the report surmised that supermarkets could do more to attract the other segments away from butchers, remarking that '... there may be certain strategies which can be put in place to create the 'feeling' or even the opportunity for familiarity without diminishing the need for anonymity' (ASI/AC Nielsen 1998, p. 53).

In addition to sharing an enthusiasm for the fresh concept, the adoption of similar retailing practices is rife among supermarkets. Thus the existence of a spectrum of differentiated retailing forms becomes a pertinent issue. In the mid-1990s, supermarket chain company documents described remarkably similar retailing principles: offering convenience at every turn (while getting consumers to choose, bag, weigh and generally do the labour performed previously by employees); shop-floor expansion (which as Kingston (1994) points out makes it longer for consumers to move around and find what they want); becoming the developer for regional shopping centres (which take a long time to travel to); local or micro-marketing (which means targeted variety at store level); and shopping linked with leisure, entertainment and family times. Given how arduous the consumers in the focus groups found shopping, the last promotional point is understandable.

The issues of convenience, family bonding and child-friendly environments become blurred in supermarket publicity. For example, the 1995 Coles Myer Report noted that '[l]ate night trading by supermarkets meets customers demand as does Sunday trading, which allows families to shop together'. Family shopping activity has a further benefit for retailers: it is the breeding ground for the junior consumer. As children wander the aisles they see, learn and demand their own commodities, creating the impetus for parents to acquiesce to children's food wants for the sake of family harmony.

The preceding fragmentary evidence indicates that Australian retail capital is being invested in cultural activities. Retail capital buys products to sell, but it also buys cultural capital in the form of extensive advertising, service delivery and family outings. This is the highly charged context which all supermarket products, including hundreds of chicken meat products, inhabit. It is debateable whether chicken would have continued its growth in consumption without the happy coincidence of supermarket chains using the 'fresh' label to differentiate themselves from others at the time when chicken became fresh again thanks to the cool chain.

THE SPECIALIST POULTRY RETAILER

Chicken is doubly blessed it seems. Not only is it a fresh, and by implication a healthy and natural food, it continues to be thought of as 'special'. Both processors and supermarkets attribute expert poulterers as responsible for reproducing chicken's status as a special food. Often the specialist poulterers are from the same migrant families who played such an important role in the mass production of chickens in Australia.

They brought with them not only small acreage farming skills but an appreciation of the versatility of chicken: they could see beyond the roasted whole bird. Several such families figure prominently in the official history of the chicken meat industry (Cain 1990), and their stories offer insights into producer-retailer relations. One family, the Moreno's, had been a successful medium sized poultry producer up until the mid-1980s, but subsequently exercised greater influence over the chicken meat complex as a poultry retailer. Something of their operation is described here.

Frank Moreno started work in the family processing company at age seventeen. He, like other members of the family, worked fourteen-hour days, starting at 5 am. In the mid-1970s his parents opened a poultry store in St. Kilda, and not long after Frank began managing their third store at the Prahran Market. This was an extension of the family's dream of having a chain of poultry shops, which began in the early sixties when a cousin had up to twenty stores around Melbourne, some of which were installed with wood-fired ovens imported from Italy. Cousin Moreno wanted to provide roasted chickens, but according to Frank 'was far too ahead of the times', with the response from housewives being 'how dare someone cook my chook'. Many of those stores had to be sold.

From the early days of the Prahran shopfront, Frank displayed breast on the bone, Maryland drumsticks and chicken fillet. He ordered fresh, not frozen, birds. Restaurants started dealing with the poulterer and women were requesting boneless cuts because, as they explained, 'Frank, that's how my husband likes his steak'.

By 1998, Moreno poultry shops were sited in several of the major, new shopping centres around Melbourne, as well as continuing a presence in the 19th century Prahran Market. Frank Moreno was particularly proud of his store in the Knox Shopping Town. There it was possible to buy free-range chickens, corn fed chickens and 'pink' or unbleached chickens. The shopper could watch as butchers portioned chickens and prepared sauces, and they could solicit advice on how to cook the available squab, duck and other lesser known poultry items which sat beside the crumbed chicken nuggets and fillets. It was a showcase of the old and the new, the healthy and less healthy, and of the entire poultry range available in Australia. In the company of game, chicken was made special once again.

On the basis of his experience, Frank Moreno argues that innovation has four sources. The first is customer demand, reflected in his case by his migrant customers asking for a range of cuts, especially legs, unlike Anglo-Australians. His own experimentation followed, driven by wanting to stay ahead of the competition. He was the first to supply fresh quails; customers then began requesting pheasants. He developed a reputation as an innovator, and game growers would approach him first with their product. The third well-spring for product development was experimentation, forced by needing to balance out the stock. 'If you have so much leg meat left over, you have to produce chicken patties, or if you have surplus legs then you have to start stuffing legs'. Finally, there is what he called 'out of left field', best illustrated by the requests for skinless chicken as a result of health fears.

Frank explains the highs and lows of being a small businessman in the supermarket era:

> I thrived in it, and had to open a new factory. I did a deal with my cousins to use an existing plant as my value-added plant, but their ability to finance it dried up. I was selling four tonnes of chicken fillet to Safeways [supermarket chain], and Inghams went in and offered the same amount for twenty cents a bird cheaper. It was a price cutting strategy to force us out.

One senses Frank's vulnerability. His Knox shop is at the entrance to a Coles supermarket and it too supplies a wide range of chicken products. He cannot compete for long if a processor like Inghams or Steggles can consistently sell chickens to this supermarket at a lower price than Moreno is able to buy them for.

Operations like the Moreno's and Lenards' were judged by a Coles National Poultry Buyer to be his major competition in the area of value-added chicken products. 'Forget the other supermarkets' he said. Supermarkets are busy imitating the Moreno-type operation because the latter caters to those with money to spend on value-added fresh food.

PRODUCT DEVELOPMENT

Although specialist poulterers may be credited with being innovators, it is the supermarkets that have the power to demand that producers turn innovation into mass product. For this reason chicken meat processors argue that supermarkets have exerted the most significant influence on product development in the last ten years. They have been the forces behind two trends: mass innovation, and coherent product ranges that operate symbolically to distinguish one chain from another. Corporate status projection has joined value-adding to products as a most important marketing device.

Each of the major supermarket chains has stores aimed at the upper, middle and lower income groups. Over the last decade Woolworths and Coles have been battling for position as major retailer to upper socio-economic status groups. They are both seeking what is referred to as the 'consumer driven' market: those who consider attributes other than price, including what are called lifestyle factors. In Britain, the Institute of Grocery Distribution has identified a similar phenomenon. The food retail market, it is said, 'will segregate into price dominant and quality dominant sectors ... The "haves" will seek innovation, excitement and high quality, the value-added products' (Harvey 1998, p. 23).

And chicken continues to play a not-so-humble role in market place manoeuvring. A national poultry buyer for Coles reported that the company purported to enlist chicken in the chain's re-positioning to appeal to higher socio-economic groups. To this end, Coles Marketing devised 'All Things Poultry', the specialty in-store poultry shop, which would compete with their revamped deli sections whose focus was also on chicken products. Rebadged 'Deli fresh', the sections came with the byline 'we will spend hours to allow you to serve in minutes'; an idea adopted from a trip to the United States by the then National Category Manager for delicatessens. The first 'All Things Poultry' store opened at Melbourne's Knox shopping centre in 1996.

However, upgrading the deli operations and introducing specialist in-house poulterers means reversing a long standing part of the Coles' organisational culture: the demand by senior management to cut the wages bill. Given that the new poultry shops aim to emulate the personalised service provided typically by butchers, each requires staff who can bone the chickens on the spot 'to give the customer the impression that it's so fresh'. More, not less, staff are needed for the in-house value-adding and product development — including the imaginative use of off-cuts: 'the drummies, wings and thighs' which are left over after the breast meat has been sold.

Supermarkets are thus operating more as secondary producers, or food manufacturers, taking less final product from their traditional sources. Coles argues that the margins with value-added are much greater than with fresh, undressed product and it is worth the extra labour costs. It is a development which means directly competing with independent poultry retailers. In some cases, the specialist poulterers are offering consultancy services to the supermarkets to help the latter ape the former.

The risks for the small businesses are enormous, so too the pressures on the National Poultry Buyers for the supermarket chains. In the case of Stan Bennett, Coles' National Poultry Buyer, the job can be likened to being in a military command post. Bennett, who had worked for the company for seventeen years at the time of interview, procured chicken meat for Coles and Bi-Lo supermarkets, Red Rooster and New Mart, a new Coles Myer acquisition in Western Australia. He enjoyed working with the chicken meat industry, describing it as 'switched-on'.

In every supermarket there is a basic product range, built up on the basis of the area's demographics and psychographics.[8] As previously identified, product ideas come from processors, from overseas, or a 'me-too' copy from KFC or a specialist poulterer. As Bennett told me, the highly competitive nature of the market means there is a persistent search for ways of increasing profit margins, which in turn incurs the seeking out of product suppliers who meet certain criteria:

> Just as customers expect good service and quality from our company so we expect those same things from our suppliers — we are a tough company to deal with ... Not only do suppliers have to provide good quality product at competitive prices, [they have to] be able to deliver consistently and on time.

Suppliers also have to produce new lines literally overnight to stay in the game. Indeed, the list of Coles' preferred suppliers is not exhaustive, a characteristic of more centralised supermarket-controlled supply chains.[9] Cost of inputs and basic safety standards are the key quality considerations for Coles in relation to chicken: what one might call, very Fordist concerns.

According to Bennett, one 'sensational line' was a lucky find. Coles asked Steggles (Queensland) to supply it with chicken stir-fry strips made out of thigh meat. Perhaps because it was supplying a lot of legs to the fast food sector, Steggles had an abundance of breast and so without informing Coles the processor was providing half-breast, half-thigh. The mix proved to be popular with consumers and Coles could not keep up with demand. Coles subsequently approached its other Queensland supplier to provide the rest: but this required the company investing in machinery which cuts breast fillet and gambling that the product would continue to be in demand — too big a risk for most small and medium processors.

But suppliers are not the only ones under duress in the quest to make profit on the back of new products. For the supermarket poultry buyers, the stakes are high. In one month, for example, Bennett had closely monitored the fortunes of the Knox shop; worked with Western Australia on its first All Things Poultry store; had cancelled a significant contract with Coles' oldest and biggest supplier, Steggles; and had sent several new products provided unsolicited from small processors to quality control for evaluation. The most stressful event, however, was the orchestration of a trial of a new supply chain system, called the 'cross-docking' of poultry, which is described below.

MARKETING AND DISTRIBUTION NETWORKS FOR THE MASS MARKET

The marketing and distribution arrangements of table chickens have long been a straightforward arrangement, with few farmers and no marketing authority being involved in distributing produce. The industry is characterised by two large retail chains and a relatively small number of processors co-operating to articulate demand and supply. Up unto 2000, when Steggles was sold to Bartter, the special relationships that were forged thirty years ago respectively between Coles and Steggles, and Woolworths and Inghams, remained. All players agree that the consumer has benefited in terms of low retail price from the extremely short distribution chain.

Despite the decades-old arrangements however, tension between the two biggest processors and the two biggest supermarket chains are marked. The Inghams' general manager for Victoria described his dealings with a national poultry buyer as a daily series of slanging matches over the phone. He likened it to a game played over and over, following the same pattern. First the supermarket buyer would demand chickens for twenty cents cheaper per bird than they paid yesterday, which would lead the processor to yell back that the supermarkets were buying chickens cheaper in the mid-1990s than they were in 1989. The processor manager would then argue that the processors have:

> ... allowed the supermarkets too much power. We're trying to say no, bugger off ... but it's hard, the poultry business is a low margin, volume driven business, we can't go anywhere else [but to the supermarkets]. If we lose the supermarkets, that drives up the cost of the rest of the product. We jealously guard that volume with the smaller operators nibbling around the edges.

A grower's representative put the balance of power more graphically when he described the preferred supplier arrangements. He believes the complex is segmented between the four major actors while 'the rest [of the processors] make like the underbelly of the whale in their dealings with supermarkets'.

From the mid-1990s supermarket company reports have stressed the importance of 'supply chain dynamics' in lowering costs, thereby advantaging consumers through delivering cheaper food. What they do not promote is the effect these dynamics have on shaping both producer operations and consumer expectations.

In considering the impact of retailers upon consumer food choice, Dawson (1995 p. 78) argues that large retailers accumulate power through a small number of management technologies. One of these is called the administered marketing channel, characterised by one party seeking to achieve economies of co-ordination of supply. Administered relationships are replacing those based on more flexible transactions, where producers and suppliers engage with one another on the basis of what products are available as well as what stocks are running low. Within the chicken meat complex the administered relationship is called the 'cross-docking' of poultry.[10]

In 1996 Stan Bennett oversaw a trial of the new system in Queensland. The system that had prevailed for many years involved processor-employed telesales staff approaching individual in-store deli managers for their daily orders. This step afforded the processor an opportunity to inform the deli manager of what was available each day and it allowed processors 'to even out' across the range of stock. Cross-docking is designed to reverse this flow of orders, with Coles' state distribution centres faxing processors with the daily order. The central order is built up on the basis of deli managers' daily assessments of short falls in stock, communicated electronically to head office. These direct sourcing arrangements leave little room for negotiation, forcing processors to deliver what is ordered rather than what they can readily make available.

The processors support the scheme because they are spared making hundreds of phone calls soliciting orders and they make deliveries to far fewer outlets. In addition, fewer contracted drivers need to be paid because the entire statewide order is dropped at the supermarket chain's central dispatch facility. Coles' company trucks, already half-full with red meat, deliver individual orders to the delicatessens, bypassing individual supermarket store departments. Coles believes that it can save wages bills for storemen and that state branches and national office can exercise greater control over the ordering and pricing decisions. The new arrangement accords with Dawson's argument that:

> [the] objective of the administered arrangement is to seek economies of coordination such that cost is removed from the channel, for example, through minimising stock holding and transport costs, reducing paperwork ... reducing the need for market searching by the retailer etc. (Dawson 1995, p. 80).

Such arrangements are less flexible than the previous 'transactionally co-ordinated marketing systems'. The new system is predicated on Coles using its traditional national processor augmented by one or two medium sized processors in each state. Having identified the cross-docking partners, all other processors will be locked out of the supply chain unless they offer very specific products. One likely impact is that the large and medium sized processors will get bigger and some small processors may sell only to butchers.

What is clear is that the new co-ordinated supply arrangements require certain investments to maintain them. Despite the long history and mutual benefits between the major supermarkets and chicken processors, these have not been easy relationships. Major tensions have arisen because chicken processor profits are clearly dependent upon supermarket demand for product. Both make money with chicken on the basis of selling huge volumes of particular portions at the cheapest price, a practice that leaves processors with large parts of unwanted bird (as well as a spur for product innovation).

To facilitate supply chain partnership arrangements, Woolworths and its traditional partner processor, Inghams, have undertaken a series of workshops to consolidate their *partnership*. The practice of partnerships replaces written contracts as the basis of horizontal integration.[11] As the strategic horizons document which provided a framework for one set of supermarket-processor workshops states:

> [p]artnering can be defined as the supplier and retailer agreeing to work closely together to leverage their combined resources. Partnering aligns strategies, systems and resources to improve mutual efficiencies whilst enhancing the offer to the consumer. A fundamental of effective partnering is the recognition that both party's profitability goals are complementary (Strategic Horizons 1995, p. 14).

Dawson (1995) points out that the ability to drive administered marketing systems creates new forms of channel power that reinforces retailer power relative to the power of suppliers and consumers. With the shift in the balance of power towards the supermarkets, the general manager of a state branch of the biggest processor conceded that 'at the end of the day we have a commitment to make Woolies competitive nationally'. The national poultry buyer for the other large supermarket chain reinforced the prevailing inter-firm dynamics with the observation that 'there is a preference for those [processors] willing to look at different ways to go forward with us into the future'. Those firms which profit from the new supply arrangements grow bigger, other firms are forced out, which in turn decreases the choice of firms for producers to sell to and so the spiral of size, profitability and concentration is perpetuated.

The cross-docking arrangements described above should save costs to both suppliers and retailers, but Foord et al. (1996) point out that some save more than others and that risks are not shared equally. The cost and risk of holding stock, for instance, is shifted away from the supermarkets onto the processors. Stockholding and warehousing are significant issues with a perishable item such as chicken and the matter of holding onto stock for considerable periods of time is compounded when supermarkets demand particular parts of the product only. None of these issues are pertinent yet to small free-range farmers.

MARKETING AND DISTRIBUTION BY THE FREE-RANGE FARMER

In contrast to the advanced logistics of large supermarkets and processors, distributing free-range product is generally a basic affair. In the case of Glenloth Poultry, described in Chapter 5, one of the four partners makes a 600 kilometre round-trip once a week dropping off to restaurants, the Queen Victoria Market and a few other inner Melbourne outlets.

Marketing free-range product, in contrast, is more complicated. The two Glenloth free-range farmers consulted for this book emphasised the absence of government labelling regulation as a constraint to the growth of their niche market. A lack of enforced labelling regarding the product's free-range status means that consumers do not know whom to trust. Contributing to this is the practice of false labelling of birds by some butchers, coupled with those same butchers' cynicism about the very existence of free-range chickens: a cynicism that is communicated to consumers, according to the Glenloth farmers and at least two focus group participants. One specialist poulterer, on hearing from his supplier that the weights of the birds were lower than usual because of the wet and cold weather, retorted 'oh, *they* don't feel the weather', and cancelled his weekly order. The only way of ensuring the purchase of a free-range chicken in Australia, I was told, is to ask for a written guarantee from the retailer or to buy a chicken that is bagged from a known free-range producer.

For the Glenloth chicken producers, their major dilemma is a lack of assured demand to make it worth investing in more shedding to increase the flock. In spite of demand for more than their 300 birds per week, they were too nervous of wavering consumer sentiment to go into debt in order to expand their operation. In 1996 the wholesale price of a free-range bird was $4.20 per kilogram, meaning a retail price of between $8 and $9, or twice that of an intensively reared bird. As the directors muttered, 'if we could sell all the birds for $3 each for the first four weeks, we'd have people hooked, and then we could increase the price and we'd get it ... but we can't afford to do that'. This was exactly the tactic used by Inghams and Steggles in the late sixties, the era of the price wars described in Chapter 5, and often nominated by chicken meat producers as a watershed in enticing Australians to consume a new, cheap meat.

LABOUR AS A FACTOR OF DISTRIBUTION

In regulation theory, labour process issues are fundamental to the nature of any product: if you rely on machines operated by unskilled labour you get a product that is very different to the one produced by skilled craftsmen. The descriptions in Chapter 5 about the different products from Glenloth Poultry and Inghams are a case in point.

This section concentrates on the management of delicatessen sections in supermarket chain operations.[12] In 'Women get their chance to run the store', the reader of *Supermarket* magazine is provided insight into the prerequisites for supermarket managers. Women, we are told, 'make good store managers because they are far better housekeepers by their nature, and pay more attention to things such as cleanliness and hygiene. They are very good at controlling budgets and are better on detail' (Mencken 1996, p. 11).

Indeed, the importance of trained and committed female staff in supermarket delis was emphasised by Coles' national poultry buyer when he described the new Knox store referred to at the beginning of the chapter:

> We've got two great ladies down there — they're dealing with a 120 foot long cabinet, and the manager is presently working fifty to sixty hours to make sure that it is working out. Casual people haven't got the care that the middle-aged have got — they need the job, rather than the teenagers who are just working for pocket money after school. I'm worried about how we are going to find another 400 girls as good as the Knox ladies to put in future poultry shops.

Coles sent the two Knox 'ladies' to Marven Poultry for training, where they spent five days between the processing and the value-added plant. They learnt not only boning techniques, but saw the lead times that a processor requires so that realistic ordering is achieved. Providing supermarket employees with extensive skills training for specific tasks is rare (Ryan & Burgess 1996) and obviously contrasts with the sort of training required by fast food chains (Reeders 1988). This establishment of in-house specialist poultry sections confirms how supermarkets are moving directly into food processing. They intend to simultaneously dictate to processors on product range developments, and to undertake their own value-adding and product development. It seems that the new in-store poultry sections constitute a post-Fordist dimension to supermarket operations.

But what does selling chicken involve for most supermarket workers? One deli manager, Anna, along with her contemporaries, has been the mainstay in selling a vast array of chicken products across Australia. Anna has been with Coles for twenty years and the changes she has witnessed to the available chicken product range and to her working conditions are instructive. When she began in delicatessens Anna sold rotisserie chicken, chicken pieces and chicken roll. Now a third of her deli case contains myriad chicken product, among them, marinated portions, chicken sausages and chicken liver pate. And while the in-store butcher next door handles chicken products, his are prepacked. Supermarkets' expansion of their delicatessen sections makes commercial sense when one considers that smallgoods 'generally realis[e] between 30 and 40 per cent gross profit' (Lee 1994b, p. 18).

For Anna, her duties consist of: choosing 'girls', training them and keeping an eye on them; processing orders; monthly stocktaking — ensuring products are available and that the shrinkage (out-of-date product) is kept to a minimum; sticking to the budget; dealing with daily memos about new products; keeping the display case full; cleaning the display case; serving customers; and putting stock away. In addition, she used to 'spit' the rotisserie chickens, but her elbows have 'gone' and she no longer does this.

Each month, Anna is given a budget from head office that contains a wages component. With the rest of the money she buys stock, adding to the core product range on the basis of her in-depth knowledge of the area and the customers. She often trials new products, under enormous pressure to report a profit. So the job has got bigger: 'I wouldn't do it again — it's a big responsibility, but it's all I know', and there are fewer staff with which to do it. Also the staff, if young, have a '[don't care] attitude — they are there for the pocket money'. Anna has to impress upon them the importance of the food's appearance, customer service and the need for cleanliness: 'I tell them if you wouldn't eat it, don't sell it'.

The shift to a casual-based workforce troubles both this deli manager and the national poultry buyer. The typical employee structure of a Coles supermarket involves two casual positions to every permanent position, with part-time staff positions falling some where in between. If the supermarkets are to sustain their fresh food offerings, their in-house processing and an attempt to create 'familiarity', clearly they need to rethink the casual/tenured labour force divide.

While Anna expressed satisfaction with the job — 'I like what I'm doing, I know I'm good at it, I know what I'm doing' — she said that the relationship with the employer had changed. 'Years ago you felt appreciated, now it's just go, go, go. The bonuses have gone, and there's too much stress. After the salami scare, and you read about salmonella — you wake up and think, Jesus, what a responsibility. It frightens me at times'. Now she is a manager, she works forty-two hours per week, with unpaid overtime and says, 'you do what has to be done'. She is not in the union, because as a manager she argues that she cannot go on strike, but believes that the union doesn't do as much as it should for the workers.[13] She continued:

> [t]o be honest I had a nervous breakdown four years ago — partly work related and all the stress of the job. It's much harder than it was, and there is less reward. It doesn't have to be monetary, just someone appreciating what you are doing, and more staff. We can really only do what is expected of us with more staff.

On my asking whether it would be easier to be the grocery manager because that job doesn't entail so many food safety risks, she answered that 'the grocery manager is treated much better than me'. I questioned whether this is because he is a man, whereupon she laughed: 'as I was saying to someone else the other day, you have to have balls to be anyone in this company'.

What about the end of the day for the deli manager and other full-time working women? Anna made the point, 'I like to cook but most nights I'm too tired to care. Luckily my husband doesn't mind cold meats, salad and bread. I buy a lot more heat-and-serve meals than I used to. It's the only way for working women'. Even poultry company directors often buy Red Rooster chooks on their way home. As one explained it to me, a whole chook allows mothers to feel that they are being an adequate provider of nutritious food and, more importantly, that they are providing a proper meal as their mothers had done twenty years earlier. The kids love it because it is 'almost' fast food. As one single father said to me in the course of being interviewed, 'chicken's half-way food. You bang it down on the table in the foil bag, and no one will complain'. This is the meal-replacement concept being pushed so hard by the supermarkets and reinforced by the KFC advertisements of the early to mid-1990s, and explored further in the next chapter.

REGULATORY POLITICS

A host of quite different government policies have an impact on food retailing in Australia. In terms of marketing the table chicken, there is only one regulation: the statutory oversighting of the dressed weight of chicken. Governments were forced to act on this matter in the 1960s when it was revealed that up to one third of the weight of frozen birds was water. In contemporary times, it is the absence of enforceable free-range product labelling that is causing concern. This, as we saw, is alleged to have limited the growth of the free-range market because consumers have been encouraged to be cynical about the existence of free-range produce. Lack of 'truth in advertising' regulation keeps some producers and retailers in the mass market.

In terms of distribution, shopping hour deregulation is of primary concern to small retailers, and one that requires far greater attention than I have been able to devote to it. In November 1996 the Shop Trading Reform Bill was passed by the Victorian Parliament. This extended the scope of the Shop Trading Act (1986) that had introduced twenty-four hour a day trading to include Sunday trading. Since that time, half of Australia has access to supermarkets day and night. A deregulated shopping hours system is probably highly significant for chicken meat consumption, because the majority of chicken products are in round-the-clock trading outlets, namely the supermarkets. In this way, red meats are disadvantaged because the bulk of that produce is bought from butcher shops trading from 9 am to 5 pm. The issue of trading hours is also significant because it further marginalises other chicken meat distributors, especially family-run stores, which cannot sustain twenty-four hour trading.[14]

Deregulated shopping hours were claimed to have been of great benefit to consumers by the recent Joint Select Committee on the Retail Sector (1999). A chicken and egg issue seems to prevail here. Traditional food shoppers, or adult women, are working outside the home requiring retailers to stay open longer. However, large numbers of women are working in retail and in food service (Humphery 1998; Van Gramberg 1999).[15] This means being available to work around-the-clock, often across family meal times. It is not clear what came first: an expanded retail and food service sector requiring women employees, or women working outside the home requiring their food related shopping and cooking to be made easier.

Absence of regulations is just as significant an issue as their presence: product labelling in a case in point. Arguably, the most important lack of regulation concerns the monitoring of written and unwritten contracts between processors and supermarkets, and in the Australian instance one could speculate collusion between the major parties. It is possible to surmise that unregulated supply chains or marketing channels have contributed to the growing concentration in both the supermarket and food processing sectors. Furthermore, the recent decision not to cap the market share of the three largest supermarket chains advantages them at the expense of independent retailers. The Parliamentary Committee which studied the issue refused to accept a recommendation from the small retailers group that total market concentration be limited to seventy-five per cent for the top three chains and twenty-five per cent for individual chains (JSC on the RS 1999, p. 8 and p. 21). This is despite the Committee being established, in part, as a response to the demise of small food shops and in spite of it recognising that their demise has underpinned the high levels of supermarket chain profitability. These are the very profits that have provided investment capital for new technologies, including the introduction of the computer-based ordering systems (which make direct sourcing from growers and processors possible). The Committee argued instead that the public good, if measured in terms of numbers employed in retailing and company share prices, warranted their decision to maintain the *status quo* (JSC on the RS 1999, p. 7 and p. 21). Of note, Australia's largest trade union, and the major union representing retail employees, argued against capping market share. The Shop Distributive and Allied Employees' Association submitted that the public interest would not be served by what it believed would follow from such a regulation: the limitation on the number of higher paying jobs in the retail industry (JSC on the RS 1999, p. 80).

Even a brief excursion into regulation in the spheres of distribution and exchange indicates that Australian supermarkets inhabit a benign regulatory environment. Further research is required to explain a lack of state interference. One reason may be the way supermarkets are in such harmony with Australia's culinary culture. Symon's food history (1982) reveals that, with the exception of the backyard chook house, white Australia has always been industrially fed, and it follows that supermarkets have become a valued social institution because they started life as the purveyors of processed foods. Unlike agriculture, they have remained until recently non-contentious in Australian politics. Ironically, it is those with some connection to agriculture who have begun to challenge supermarket dominance over agrifood supply chains.

CONCLUSION

This chapter adds to the unfolding explanation for chicken's popularity. Chicken's acclaim began with its status as a familiar yet rarely eaten food in the Australian culinary culture. This *special* food quickly became cheap and readily available, thanks to those in distribution as surely as those in the sphere of production. The 1960s rapid expansion of two national retailing networks, which had entered into arrangements with the two national processors, streamlined marketing arrangements and obviated the need for a statutory marketing authority. Moreover, corporate decisions regarding the acceptability of low profit margins per bird led to consistently low wholesale and retail prices for chicken *vis a vis* other meats. Low wholesale prices arguably facilitated the spread of take away chicken outlets across the country, as did the excitement generated by the arrival of Kentucky Fried Chicken.

At this point, food retailing in Australia appears to be in a period of transition from a Fordist regime to a mixed regime containing post-Fordist elements. On the one hand, the infinite product development for the new in-store poultry sections, combined with the specialist training being given to the staff in this area, suggests post-Fordism. Moreover, a retail labour market, characterised by numerical flexibility, is evolving to cater for deregulated shopping hours. In addition, the new cross-docking supply arrangements will introduce Just In Time delivery systems: another post-Fordist feature. On the other hand, the case for flexible systems is diminished because the new supply arrangements are more centralised than the transactionally co-ordinated marketing mechanisms they are replacing. Furthermore, what we see in the more traditional deli sections, as opposed to the new in-store poultry sections, is a continuation of the routinisation of the labour process. What we are witnessing is the emergence of two retailing forms under the one supermarket roof, but it is hard to discern truly alternative retail systems offering different forms of service and methods of exchange.

This chapter also offers more complex insights into the changing balance of power in the chicken meat commodity complex than that afforded by regulation theory. It shows that it is premature to talk of the demise of modern supermarket based agribusiness, as Sokolov (1991) has done about the United States. In relation to chicken meat, the balance of power within the distribution sphere has shifted in the last decade away from processors towards supermarkets. The reasons include the latter's ready access to large amounts of capital, lack of state regulation of supermarket activities and their superior market intelligence and cultural activities.

The Australian chicken meat case study certainly lends support to Hughes' assertion that a lenient state allows retail capital to assume a hegemonic position in relation to other forms of capital (Hughes 1996, p. 96). This accumulation harbours the ability of supermarkets to oversee the restructuring of supply chains. Because of their shared history, the reconfiguring that is taking place in the chicken processor- supermarket supply chain is less dramatic than that happening in fruit and vegetable supply chains, where wholesale markets are disappearing (Parsons 1996). However, its one-sidedness, despite the talk of partnership arrangements, is noteworthy. Of all commodity groups the chicken farmers are the best organised in the country and the concentration amongst the processors is possibly the highest of any agrifood commodity. Yet as evidenced by the new supply chain arrangements, the supermarkets are in control. While chicken meat producers have so far successfully repudiated global restructuring in the form of free trade and are prevaricating about the need for labour market deregulation (see Dixon & Burgess 1998), they are succumbing to retailer-led restructuring.

Clearly retail capital is not confined to the circuit of realisation, but is moving into the circuit of production. It is doing this both directly, through investing in on-site manufacturing of products, and indirectly, through demanding that processors supply certain products. In the case of chicken meat, supermarkets are particularly mindful of the competition posed by specialist poulterers for consumers with high disposable income. Some large retailers are reinvesting retail capital in commodity production activities to compete with these niche market suppliers, and in so doing we are witnessing the emergence of product differentiated retail systems, as Harvey (1998) has identified for Britain. Because products are valued by the large retailers for the way they represent particular values and for the way they contribute to the corporation's cultural capital, product ranges, to use a Bourdieuian concept, have symbolic power. Through their product portfolios, supermarkets communicate what the corporation stands for and what regimes of value they support, secure in the knowledge that market place positioning is not only important in attracting consumers but for attracting share market investors.

In short, supermarkets are emerging as the most powerful actors in the distribution sphere and are exerting considerable influence over the production sphere of the chicken meat commodity complex. Supermarkets are placed by their economic geography and their specific food-related activities at a number of intersections: processed food and fresh food distributors; job creators and family business destroyers; mediators between producers and consumers; facilitators of differentiated communities of consumption; and intervening in the rural-urban dynamic in terms of their place in Australian suburban life and in their relations with agriculture. To remain non-contentious in the social landscape requires managing these contradictions as surely as managing the supply chain.

Put another way, supermarkets have some precarious 'social risks' to manage, just as the chicken processor has 'perishable product' risks. The preceding remarks underscore the strategic importance of mobilising bias in regard to what is good to eat, as well as corporate retail interests. Retail trade magazines contain numerous examples showing how KFC and the supermarket chains continue to sell a way of life, as Humphery (1998) characterised supermarkets of the 1960s. Selling a way of life is multifaceted and involves particular language, images and practices. The consolidation of particular regimes of value, through the production of food discourses and practices, is the subject of the next chapter.

7

DISCURSIVE PRACTICES OF THE CHICKEN

QUESTION:
Why did the chicken cross the road?

ANSWER:
To catch the tram.

In early 1996 my husband came home and told the following story. A colleague arrived back at her workplace after lunching with a friend in the city. The friend had told of how, a few days earlier, she had caught a tram outside the Queen Victoria Market, Melbourne's famous fresh produce mart. What she witnessed had shocked her. A woman of European descent stepped onto the tram with a live chicken under her arm.[1] The tram driver told her that live animals could not travel on the tram. She muttered something, got off the tram, and appeared again a few seconds later with a dead chook under her arm. My husband and his colleague, and my husband and I several hours later, marvelled at how fantastic this was, without really elaborating why.

Two weeks later, while reading the book *Tucker in Australia*, written almost twenty years earlier, I came across the following passage:

> [i]n the good old days it was possible and perhaps advantageous to buy second-hand spectacles and false teeth at the market. One popular story tells of the lady who bought a live chicken at the Victoria market, then boarded a city tram. When the trammie asked her not to bring live animals on to the tram she promptly alighted, wrung the bird's neck and reboarded the tram (Wood 1977, p. 76).

Was the colleague's friend telling an actual experience or a story she had heard? More importantly if it is the latter, why is this particular story still in circulation? Why is it so appealing? Is it the wonderment at rare skills: of people assured, rather than terrified, of their food supply, or is it bemusement at the persistence of peasant traditions? Is it a story peddled by the near-by traders wanting to attract customers?

Appadurai argues that myths arise as the spheres of production, distribution and consumption become disarticulated. Furthermore, 'it might be said that as the spatial and institutional journeys of commodities grow more complex, and the alienation of producers, traders, and consumers from one another increases, culturally formed mythologies about commodity flow are likely to emerge' (Appadurai 1986, p. 48).

The two Glenloth directors, quoted in previous chapters about their free-range farming of chickens, despair at consumer and urban ignorance of agriculture and what is involved in food production. One said:

> [g]o back to the fifties and earlier, twenty-five percent of people had contact with the land — farming it or helping out. City people had relations in the country and they identified with it. There have been enormous changes in my lifetime, and there is enormous lack of understanding on both sides. Neither understands the problems of the other, and there are great pressures on agriculture.

According to several producers whom I interviewed, consumers have unrealistic demands when it comes to food supply. They spurn animal cruelty but they are not prepared to pay the oncosts to eliminate it from their food. Likewise, consumers do not want chemical additives but are not willing to pay more for organically grown food. Possibly the consumer contradiction about not wanting to pay more to have their food supply fears allayed explains why the focus group consumers 'don't want to know the full story' about chicken meat production. They consider poultry production to be an offensive industry, 'wanting it out of sight, out of mind'. Can consumer resistance to spend more on food coupled with ignorance explain the evidence from Chapter 5 that producers are indifferent about improving the quality of their products? Or does the history of producer success, without engaging directly with consumers, encourage them to continue to adhere to Henry Ford's aphorism that, 'consumers can have any colour as long as it's black'. The Fordist delivery of standard products, not quality products, appears to suit the majority of producers and consumers.

In this chapter I take the explanation for Australia's most successful post-war agrifood industry one step further by explaining how contradictory consumer beliefs and behaviours are perpetuated. I specifically explore the busy intersection of cultural and economic activity by focusing on the networks of retailers, advertisers and other food knowledge producers who have been producing and circulating understandings and mythologies around what constitutes *good* food and *desirable* food practices. The first part of the chapter uses secondary sources, augmented by material from interviews with producers and retailers, to examine the efforts that have been made to promote chicken as *good to think*.

In the course of this particular line of research, I became aware that chicken meat was not only an item being promoted as a lifestyle choice but was being used simultaneously to sell corporate identities and particular ways of life. Accordingly, I have included material that illustrates how chicken meat has been used by KFC to reconfigure its image. I also look at the way chicken meat is being used to reshape understandings of home-life and meals through the home-meal-replacement strategy. This is a strategy designed to influence consumer lifestyles and, I believe, lies at the heart of the major food retailers' capital accumulation strategies: in particular the devaluing of home cooked food and the revaluing of food cooked in commercial kitchens. The new material in this chapter sheds further light on the distributor-centred networks, leading me to identify a particular configuration of actors that constitutes what I am calling the producer-consumer services sector. The chapter serves to reinforce how important cultural production on behalf of major corporations is to the economic processes underpinning the balance of power in the chicken meat commodity complex.

KFC

Because KFC figures so prominently in what follows it is necessary to provide a snapshot of its history and describe the way the parent company, PepsiCo, boasts of its Australian operation. According to company documents (PepsiCo Restaurants International Fact Sheets n.d.) the KFC story began at the turn of last century, when a young boy, Harland Sanders, became an accomplished cook through 'family necessity'. He spent considerable years doing casual work and serving in the United States Army, where he received the title 'Colonel'. Or do we have another myth? *The New Economics of Fast Food* says that 'the honorary title was bestowed upon him by Kentucky Governor Ruby Laffoon as a tribute to Sanders' contribution to culinary arts' (Emerson 1990, p. 7).

Anyway, at age forty, Colonel Sanders purchased a service station, motel and cafe in a small town in Kentucky. Over the next ten years he tried different seasonings to flavour his chicken. From this experimentation evolved 'his secret recipe of 11 herbs and spices and the basic cooking technique which is still used today'. The KFC official history indicates that he sold the business when the town was bypassed by a highway. He then 'travelled the United States by car, cooking chicken for restaurant owners and their employees. If the reaction was favourable Sanders entered into a handshake agreement on a deal which stipulated a payment to him of a nickel for each chicken the restaurant sold'. By the age of sixty-five the Colonel had 600 Kentucky Fried Chicken franchise outlets dotted across the United States and Canada. This was 1964, the year in which he sold the American business for $2 million, leading to another rags to riches story, or as corporation documents express it, '[t]he 65-year-old gentleman had started a worldwide empire using his $105 social security cheque' (PepsiCo 'KFC History Colonel Sanders', n.d.).

Australia's first KFC store was built in Guildford, in Sydney's western suburbs in 1968 and within eighteen months a further twenty restaurants had been built, thus sparking 'the fast food revolution in Australia' (PepsiCo Fact Sheets). By 1995 the Australian network of 452 restaurants employed 12 000 staff, eighty per cent of whom were under twenty-one. As a result of their efforts, Australia contributed thirty-five percent of KFC's earnings outside the United States in that year (Shoebridge 1996, p. 65). While this constitutes a drop in contribution compared to the 1980s, Australia remains one of the most successful KFC divisions in the world in terms of its sales and profit contribution (Shoebridge 1996, p. 65).

Two-thirds of the Australian restaurants are company owned while the rest are owned and run by franchisees and all stores worldwide report results back to PepsiCo headquarters in New York. Australia is the parent company's South Pacific headquarters, covering the nations of New Zealand, South Africa, Papua New Guinea, Fiji, French Polynesia, New Caledonia, Western Samoa, Vanuatu, Tonga and the Solomon Islands. The South Pacific region is claimed to be 'often at the forefront of new developments in KFC's strategy for global success'.[2] (PepsiCo General Briefing Fact Sheet n.d.). This claim is evaluated below when telling the stories of recent KFC product development. What is indisputable is that the volume of take away chicken grew exponentially with the entry into Australia in 1968 of Kentucky Fried Chicken (Larkin 1991), and at the turn of this century the company was dominating the away-from-home cooked chicken market.

COMMODITY PROMOTIONAL ACTIVITY

Most of the attributes of chicken which were so valued by the focus group participants — low price, variety and health benefits *vis a vis* red meat — are properties not inherent in chicken meat. Rather these values have been added through advertising, third party associations with reputable bodies and public health campaigns. Encouraging positive judgements about a commodity's goodness is the goal of each of these types of promotional activity. In order to understand the genesis of the information and images of chicken that surround us every day, I provide examples of some major promotional activities. The vignettes confirm how propitious the advent of what nutritionists call 'the low-fat era' (Santich 1995b) has been for chicken.

Except for a six week national advertising campaign in 1982, the poultry industry has sponsored only one 'substantial' campaign featuring chicken (Fairbrother 1988, p. 462). In 1987, and amid the backdrop of a debate linking fatty food to poor health, the peak producer body, the Australian Chicken Meat Federation, employed a public relations firm and a well-known consultant nutritionist to stress chicken's low-fat status. The highest circulation women's magazine at the time, *New Idea*, contained an eight page insert 'healthy eating featuring chicken'. The nutritionist began her editorial:

> ... some people find nutrition confusing. However, among qualified nutritionists there is now consensus of opinion that the ideal daily diet fits the approach of the healthy Diet Pyramid devised by the Australian Nutrition Foundation — basically we need to eat ... moderate quantities of fish, poultry [and] very lean meat (Stanton 1987, front page).

The pyramid followed, as did 'six nutritious recipes' featuring chicken. An accompanying table showed the amounts of fat in a range of meats, pointing to chicken's relatively low-fat status.[3] While an industry-wide campaign has not been repeated, individual companies have on occasion advertised their own products. In 1989 for instance, Goodman Fielder promoted what was called the Steggles Champion Breed chicken, highlighting its lower fat content and claiming a twenty per cent increase in breast fillet over other breeds. The company had the Australian Consumers Association (ACA) substantiate its claims and the *Choice* article which reported the tests concluded that 'Steggles has certainly produced a bird worth crowing about! It is significantly lower in fat than normal chickens and has more breast meat as well. However, any chicken (if cooked properly) can form the basis of a healthy meal, made healthier by removing all skin and visible fat' (Australian Consumers Association 1990, p. 37). Importantly, the consumer body pointed out, as had the nutritionist in the *New Idea* insert, that a low-fat diet could be achieved by choosing most lean cuts of meat and grilling rather than frying them.

The Steggles campaign was not deemed to be worth repeating by other processors, because a majority of chicken meat is sold unbagged or unlabelled and, as *Supermarket* magazine remarked, it does not make much sense to brand chicken (Lee 1994a, p. 30). A lack of branding by producers suits the traders because they can impose their own brand, such as All Things Poultry, on the generic item. The unbranded nature of chicken has arguably contributed to the balance of power tipping in the direction of the retailers and away from the producers. People recognise that they have eaten a Woolies chook not an Inghams chicken.[4]

The most significant direct advertising of chicken has been undertaken by KFC, which admits to spending five per cent of sales dollars annually on advertising in Australia. It advertises on television fifty-two weeks of the year and also runs an average of fifteen public events per year to keep the company at the fore of public consciousness. It justifies its advertising budget thus:

> [t]here is so much choice and competition for the discretionary dollars, that all activity must be new and consumers must be continually told what is available — you must remain top of the consumers mind, as impulse purchasing is common in the fast food market (PepsiCo TenderRoast Launch n.d., p. 7).

In addition, the prominence of their 400-odd stores through signage and siting arrangements adds significantly to their presence in a range of landscapes.

A close examination of KFC television ads reveals, however, that the corporation promotes four things simultaneously: KFC products, the KFC company, particular lifestyles, and, less obviously, chicken. KFC associates its products with the qualities of the corporation (in my opinion the qualities are similar to McDonald's: dependable and successful overlayed, in the KFC case, with 'old-fashioned' values represented by the Colonel). Indeed, KFC blatantly blurs the distinction between advertising that gives facts about the product and advertising by association. Arguably, chicken has benefited by the association with this hugely successful corporation and in turn, the corporation has been a beneficiary of chicken's association with health. One in four Australians visits a KFC every fortnight to, in the main, buy fried chicken and I would think that, in the same fortnight, many of these people are also buying chicken carrying the National Heart Foundation (NHF) Pick the Tick logo.

In 1993 the CSIRO conducted a NHF sponsored consumer survey which found that foods which had some health authority approval were favoured over those that had no such endorsement. Concurrently, NHF approved products received high level support from women as one of those authorities. Forty per cent of the respondents said that they looked for the Pick the Tick logo when shopping, while one quarter said that the Tick was used to help them choose meat. It is in this context that chicken meat producers are significant users of the Tick program (CSIRO 1994, p. 43). Unlike the self-interested promotions of retailers or producers, legitimisation by what is perceived to be a medical body carries more authority (CSIRO 1994, p. 15).

To obtain the NHF logo a company must first have its product tested by the NHF and, having then met the guidelines, must pay a fee to the NHF to use the logo in product promotion.[5] The fee operates on a sliding scale, so a large company with a raft of NHF approved products may pay $100 000. This is alleged to be the amount that Goodman Fielder, owner of Steggles, paid in 1996. It is indeed a money earner for the not-for-profit research and health promotion organisation and these payments augment the bequests and donations by the public, making the NHF Australia's fifth largest charity.

The use of registered charities to promote food would make a fascinating study. Suffice to say that not all are impressed with this sort of approach to making food *good to think*. Wright (1991), a food industry lawyer who reviewed the operations of the Pick the Tick program by interviewing producers, health bodies and government authorities, found many were concerned that the scheme was likely to mislead or deceive: that the logo authorised certain products and not others, which could be equally or more beneficial to health. A few years later, nutrition researchers concluded the Tick was 'particularly misleading as a guide to healthy foods' (Scott & Worsley 1994, p. 27).

Such concerns have not dissuaded the larger poulterers from badging their products with the Tick. How consumers reconcile the sight of an processor's truck, covered in the Pick the Tick logo, at the side door of a Red Rooster fast food outlet making deliveries of Heart Foundation endorsed chickens is not known. However, from the producers' point of view the twin association of health and fast food makes good commercial sense especially as they appear to operate to the advantage of children. In a discussion of 'eight to twelve year old' consumers, the point was made that 'fast-food chains such as McDonald's, Pizza Hut and Kentucky Fried Chicken are still popular and children use the nutritional claims made in fast-food advertising to convince their parents they should be allowed to eat it' (Shoebridge 1993, p. 166).

Linking human health and corporate health extends beyond supermarkets' use of the chicken to symbolise market strength and vision, as described in the last chapter. In 1997 Eatmore Poultry created the nation's first Poultry Education website on the National Heart Foundation webpages. The site gives a brief history of the poultry industry and of the Eatmore company, followed by a thorough description of the poultry production process. While it does not address the contentious issue of hormone supplements that so troubles consumers, it is the only attempt that I know of by a producer to inform consumers of chicken meat production. The site closes with images of the NHF logo, the Pick The Tick symbol and some recipes featuring chicken. In this way a non-government health authority is cashing-in on the table chicken.

Within the context of government sponsored nutrition campaigns, it is likely that Australian chicken meat producers and retailers will continue to take every opportunity to champion chicken's 'health' promoting properties. Like its American counterpart, the Australian government has endorsed a food pyramid in which certain foods are privileged over others. The image of chicken figures prominently in the pyramid segment signifying that moderate weekly intake is nutritionally acceptable. Some are aware that being in this segment is not as ideal as being in the segment that represents daily intakes. In relation to government sanctioned dietary guidelines in the United States, the world's most powerful poulterer has opined:

> [f]or the past decade, the government has played an increasing role in educating consumers on the importance of nutrition. Depending on how you look at it, it can be a good or bad news story. In the past, the government placed equal emphasis on the food groups, but it now favours grains, fruit and vegetables. The Food Guide Pyramid and an emphasis on less protein consumption cannot be ignored in considering the outlook for poultry ... Rather than fight it, our industry needs to continue looking at ways to make it easier for consumers, who want to change their dietary behaviour, to use chicken as an ingredient in their foods ... the most obvious way to increase consumption is to be sure to sell chicken at all places where the consumer purchases food (Wray 1995, p. 15).[6]

Without a doubt, the recent 'demonisation of fat' within a context of the more sustained nutritionalisation of the food supply has benefited the chicken meat commodity complex more than any of the red meat complexes. Moreover, the white meat industry's lack of self-promotion may have been to its benefit. Unlike the 'self-interested' Australian Meat and Livestock Corporation talking up the goodness of red meat (Shoebridge 1995), chicken meat producers have let other more credible parties do the talking for them.

There is one final association of which I became aware during the research: chicken can still be considered special despite being cheap. During the last decade this association has come from a different quarter: endorsement by celebrity gastronomes and chefs. Specialist producers and retailers credit famous cooks and culinary authorities with much of their success. For instance, Frank Moreno told me that when specialist poulterers are written about in the Epicure pages of the Melbourne broadsheet, *The Age*, his sales climb rapidly. He said that on the occasion of a cookery writer describing his barbecue chicken spare ribs he had to double the volume to meet demand. He also mentioned the time when *Vogue* magazine's renowned gastronome, Diane Holuigue, called his the best poultry stores in Australia — encouraging him to expand his business considerably because of the increased patronage of his stores. Similarly the Glenloth company credits Stephanie Alexander, one of Australia's best known restaurateurs and food writers, with contributing to its success. I was told that, 'Stephanie has been great. We use her as our tasting board. The secret of a good tasting chicken is what it has been fed, and we try different feeds and then ask Stephanie to taste the results. She also promotes us whenever she can'. The latter claim is borne out by the beautiful description she gives of the Glenloth farm and operation in one of her many books, *Stephanie's Australia*, and her featuring of the farm's produce in her television series 'The Shared Table'.

The significance of being associated with celebrity chefs is that it makes chicken special again for those with disposable incomes to spend on food and eating out. It is in this way that cosmopolitans contribute to the culinary revolution that is said to be sweeping the cities of rich nations.

CORPORATE IDENTITY STRATEGY, OR THE IMAGE CHAIN AT WORK

I have previously described how supermarkets use chicken not only as a loss-leader and as a profitable commodity but as a symbol for positioning themselves. Chapter 3 describes how chicken was used to place supermarkets at the cutting edge of progress brought about by technology and more recently as offering opportunities for a fresh and healthy life.

The most audacious example of chicken meat being used to position a company comes with Kentucky Fried Chicken's rebirth as KFC in 1993; the primary purpose of which was to lose its image as the 'fried food king'. Once again, both the 1980s lipophobia movement and the broad consumer movement can claim the credit. In an Australian Consumers Association (ACA) survey of fast food, the fat content of some well-known take away meals was revealed. The ACA showed that eating a meal of Kentucky Fried Chicken's original recipe chicken and chips and coleslaw meant consuming nine teaspoons of fat. They concluded that this particular meal combination was the fattiest of all take away meals except for fish and chips and pie and chips. They suggested that eating a fast food meal more than once a week would distort recommended dietary requirements (Australian Consumers Association 1994).

Within this context, and as part of its 25th Anniversary Celebrations, KFC (Australia) announced in 1993 that it was going to spend $60 million on building and staffing new restaurants and rebuilding and refurbishing old ones. More importantly, it announced that it would 're-image' all KFC restaurants around Australia and launch a major new product called TenderRoast. The company proclaimed that this product had been developed in Australia and would be exported to KFC worldwide. At the product launch journalists were given background notes pointing out that in 1993, consumer research had highlighted that 'customers were looking for a non-fried chicken product from KFC. To keep up with the trends emerging KFC needed to broaden its product offering eg., health trends, need for variety and added convenience'. In addition, 'market analysis clearly showed the business potential available in the BBQ chicken market ... [a] market ... estimated to be $1 billion per year' (PepsiCo TenderRoast Launch n.d., p. 2).

The KFC technical group, in conjunction with the marketing group, worked closely with suppliers including Inghams and Steggles, the marinade and sprinkle suppliers, McCormicks Foods, and equipment and packaging suppliers. Consumer research involving taste panels was conducted and repeated until a product was ready for launch on a test market. The publicity blurb maintained that 'TenderRoast is viewed as a product, along with side items which provide a perfect meal replacement for the family. It is vital that we have mum's approval, however, it is equally important that the product appeals/is liked by the rest of the family' (PepsiCo TenderRoast Launch n.d., p. 5).

The launch of the product went hand in hand with a company repositioning exercise because as the background papers pointed out, ten years earlier Kentucky Fried Chicken had unsuccessfully tried a barbecue chicken:

> The product was not unique enough and the consumer gave Kentucky Fried Chicken (the fried chicken kings) little latitude in their minds to accept a wider positioning. Internally the product was difficult to cook and generally speaking a pain in the bum. So it failed (PepsiCo 'How KFC Increased ...' n.d., p. 1).

This experience taught the company that it must adjust its image before consumers would accept new products, and in turn the re-imaging had to be accompanied by a product range that included non-fried products. Two advertising agencies were commissioned: one to concentrate on the corporate name change and how the product fitted with the new developments, and one to focus on the product itself. The resulting corporate slogan was 'TODAY'S KFC — I LIKE IT LIKE THAT'. The image, so said the company blurb, is modern and contemporary, in tune with the needs and fast paced life of today's consumer. Furthermore, those attending the launch were told that TenderRoast is part of the 'NEW KFC', and the most important product launch in KFC's twenty-five years of operations in Australia. 'TenderRoast will help change the way consumers see the old Kentucky Fried Chicken'. KFC Marketing hired an advertising agency to develop the launch strategy: 'KFC have the best tasting BBQ chicken in town'.[7] This message was supported by two television commercials, extensive in-restaurant point of sale promotions and a double page spread plus value coupons inserted in *New Idea*. The same coupon and brochure for TenderRoast was delivered to every household letterbox in Australia. Public relations activities included capital city radio station DJs being delivered TenderRoast, and commenting on-air about the taste sensation.

Despite the promising campaign, research showed that further growth was inhibited by KFC's historic image as the 'fried chicken guy'. Celebrity endorsement was chosen to address this problem. After some research, and in spite of the fact that she didn't fit the typical celebrity profile for the product, Elle MacPherson, the Australian supermodel, was asked to endorse the food.

MacPherson is not a renowned chicken eater, her authority on marination techniques is scarcely known nor does she exude the image of KFC. However, the strategy was to impact on people:

> ... who are not KFC eaters, for whom TenderRoast could well be an appealing product but so far had not distinguished TenderRoast from their overall apathy to the chain ... By using Elle's health, vitality and business success we aimed to lift the profile of TenderRoast and confront those people with negative perceptions to KFC in general (PepsiCo 'How KFC Increased ...' n.d., p. 4).

Overall sales of TenderRoast exceeded fifty percent during the campaign and fried chicken sales increased as well. So the marketing strategy appeared to work in the short term and while it achieved a seventeen per cent share of the barbecue chicken market in its first nine months, in an equally short time it had dropped to nine per cent. This is a very different picture to the company's majority share of the fried chicken market.

Within two years of the multimillion-dollar launch, TenderRoast was dropped. I was told that customers did not like the flavour because 'the marinade was too strong'. One marketing analyst surmised that the failure was due to 'the fact that it was promoted as a product, not as a meal' (Shoebridge 1996, p. 66). In response Kentucky BarBQ was introduced in October 1995 and the company reported on its launch in much the same way as its predecessor, overlooking the hype surrounding the TenderRoast promotion. The material states that 'Kentucky BBQ Chicken ... is probably the most important product launch in KFC's 28 years of operations in Australia' (PepsiCo Kentucky BBQ Launch n.d, p. 4).

Like the product it replaced, Kentucky BarBQ played a role in repositioning the company. Where TenderRoast was intended to reposition a fried food purveyor as a healthy food provider, Kentucky BarBQ was used to move the company from snack or treat provider to meal provider, and particularly of family meals. To complement the non-fried meals, KFC introduced the notion of breakfast, lunch and dinner occasions, not simply the snacking/grazing/all day option. In explaining the new product range to *Business Review Weekly*, the Regional Director (Marketing) said: 'people are looking for real food fast, not fast food. They do not consider fried chicken an acceptable alternative to a home-cooked meal, but barbecued chicken — sold as a complete meal, with side dishes such as potatoes and peas — is acceptable' (Shoebridge 1996, p. 64).

Within the marketing pages of broadsheets and business magazines KFC is referred to as an image chain. The person who watches one of the KFC ads or consumes one of their meals becomes part of the image. The television commercials for the dinner meal combinations, one of which centres on the new Kentucky BarBQ product, contain 'the KFC family' who will be used to promote all of their new products. Not only is the corporation promoting some of its meals as healthy, it is also promoting them as family friendly. They are promoting a particular corporate image in the expectation that the products will be judged positively, and when the products are well regarded, so will be the company.

The product launches show two things: the extent to which cultural production is harnessed for capital accumulation, and the symbiosis between products and corporate identities. KFC is clearly pursuing symbolic power through economic power.

SELLING A WAY OF LIFE VIA HOME MEAL REPLACEMENT

At the beginning of 1996, as part of the continuation of its repositioning strategy, KFC's Manager of National Marketing explained that the movement into meals considered by women to be healthy was 'part of a movement towards capturing more of the meal replacement market rather than being simply an occasional-but-unhealthy treat' (Strickland 1996, p. 29). He highlighted that KFC's new range of products was intended not only for corporate identity purposes, but to help it compete with the supermarkets for the home-meal-replacement market. Elsewhere he added '[e]ssential to convincing Mums to replace regular meals with fast food ... was alleviating the associated guilt ... The emphasis now is on the meal, not the product'. For:

> [a]t the end of the day mum works these days and she also normally cooks. We know Mum doesn't want to replace meals during the week with fast food: she wants real food fast rather than fast food. If we just sell her chicken we're not solving her problem, so we're trying to sell her a complete meal which she feels happy buying (Strickland 1996, p. 29).[8]

One of Australia's leading processors confirmed how acceptance of these products is conditional upon the family cook expending some labour, even if this amounted to heating the food in the microwave. Then, as he said, 'the housewife can get her jollies off thinking she's cooking, while she's actually doing very little'. In these statements we see reflected both contempt for women consumers as well as a need to have their approval, a tension that is inherent in commodifying every aspect of the food supply.

As KFC was repositioning so too were the supermarkets. I was told that the Coles Myer Research Division had been researching meal replacement as a key strategy to help it woo the 'lifestyle consumer'. *Restaurant Business,* an American-based trade magazine, is a particularly rich source for tracking the emerging trend of buying someone else's meals on a regular basis. In one issue, the various dimensions to home-meal-replacement are highlighted thus:

> Is it a potentially huge new food service category, or just a fancy name for take-out? Is it strictly home-style food like Mom used to cook? Or do pasta primavera and kung pao count if you take them home? Do individual meals qualify, or are we talking strictly family style here? And do you have to eat it just at home or is it still home-meal replacement if you eat take-out chicken and mashed potatoes at your desk at work — or even right in the restaurant? (Casper 1996, p. 165).

The writer of that passage is struggling with discursive issues as much as practical ones. Meal replacement strategies do not simply mean product innovation and targeted marketing; they actually involve redefining what is considered to be *home, meal* and *replacement.* An interview with the man credited with being the creator of the term 'home meal replacement' reveals more about the importance of language:

> [w]e wrestle with questions such as what is the 'home' in home meal replacement. I think we really ought to hyphenate 'home' and 'meal', because you don't have to be at home to consume it ... That's why we've changed our vernacular to quick quality vs home-meal replacement (Casper 1996, p. 165).

This former CEO of PepsiCo points out that the hair-splitting is worth a lot of money: '[g]eneral industry estimates range from $20–100 billion depending on how you define each of those three words: home, meal and replacement'. His other rationales for the new culinary trend concerned consumer demand: '[w]e're in a revolution regarding the way consumers feed themselves in this country ... We're finding more and more people who haven't taken much time to learn how to cook, or don't like to, or else just don't have the time, and are looking for other meal solutions' (Casper 1996, p. 171). The corporate rationale for seeking alternatives to home cooking stands in contrast to the consumer perspective offered by the focus groups. Those consumers expressed ambivalence for cooking in the context of experiencing a general pressure to do the right thing but they also rated cooking more highly than any other part of food provisioning. Instead of not cooking, they want the context in which cooking takes place to change: they desire greater certainty around what constitutes good food, they want to feel less pressured by children and dinner time routines and they want food shopping to be easier. Home-meal-replacement will meet only this last need.

This fact will not deter what appears to be a sustained commercial strategy on behalf of food retailers. Since the mid-1990s Australian supermarket chains have signalled that they are competing with KFC and major chicken fast food chains for the home-meal-replacement market (Shoebridge 1996, p. 65). The market place contest will have ramifications for farming and agricultural production as well as for home life. In the *Australian Farm Journal,* under the title 'Making fresh food faster', the Woolworths' buyer of fresh fruit and vegetables highlighted that 'the horticulture industry must pitch itself into direct competition with the booming fast food sector' because:

> [t]he big fast-food outlets such as McDonald's and KFC are encouraging people to dine out more, which means consumers buy less fresh food to cook at home. The challenge for the fresh produce sector, notably growers and retailers, is to persuade consumers to eat more at-home meals ... But that won't happen unless fresh produce outlets can offer a wider range of innovative, consistent-quality, easy-to-prepare meal options (Carson 1995, p. 44).

This argument was supported by the Director of the Australian Horticulture Corporation who proposed that the horticulture industry 'has to regard itself as part of the food industry rather than the produce industry. The food industry is similar to the fashion industry with people always seeking new flavours and tastes' (Carson 1995, p. 45). Of note, these particular agrifood producers are siding with supermarkets and fast food traders in discouraging home cooking from scratch and in endorsing pre-prepared, heat-and-serve meals. Together, producers and retailers are signalling their desire to further process food in the factory or industrial kitchen rather than have this labour performed in the household kitchen.

KEY ACTORS: PRODUCER-CONSUMER SERVICES

The preceding examples of cultural production activities support the proposition, discussed in Chapter 3, that the food system is being shaped by the entry of new food authorities who can mediate a desire for foods that promise healthy convenience and family harmony. The experts in the food system are located increasingly in distribution and exchange and it is from this sphere that they contribute to aligning the activities of production and consumption in a culinary culture that is disembedded from everyday experience. The expertise is not that of direct experience of production or consumption but is in the form of having information and direct experience represented visually and symbolically. The traders in information and symbolism most pertinent to weaving stories around chicken meat consist of nutritionists, medical research organisations, KFC employed advertisers, supermarket chains, consumer groups and celebrity chefs. Importantly, they are not separate entities but rather networked actors who value-add to one another's efforts.

How can such a mix of actors, who are so pivotal to the matrix of accumulation in Australia's food system, best be characterised? Saskia Sassen in *The Global City* provides a useful entry point. She introduces the book by claiming that she is not investigating 'formal' or corporate power, but is interested in describing the economic activities that lead corporations and banks to achieve power over other entities (Sassen 1991, pp. 6–7). By adopting the concept of production sites, Sassen uncovers the practice of financial and other key services concentrating in particular cities so that they may service large corporations. Sassen argues that these 'producer services' facilitate the accumulation of corporate power because they provide corporations with capabilities to produce commodities that circulate the globe.

Amongst the oldest examples of producer services are accounting and advertising firms, but in the 21st century the list is extensive and includes finance, innovation, design, transport, communications and security service firms. Accordingly, '[p]roducer services can be seen as part of the supply capacity of an economy' representing a mechanism that 'organises and adjudicates economic exchange for a fee ... They are part of a broader intermediary economy' (Sassen 1991, p. 90).

Just as Leopold (1985) has argued about the importance of product differentiation to the prosperity of agrifood producers and Mark Harvey (1998) has linked the future fortunes of supermarket chains to product differentiation, Sassen describes how product differentiation drives the need for producer services:

> Product differentiation and the resultant market differentiation emerge as yet another set of specialized conditions that must be brought together at the higher levels of a corporation. Greater product differentiation expands the marketing and selling functions of a firm (Sassen 1991, p. 97).

Specifically, the case of chicken meat reveals the importance of nutritional producer services to product differentiation. Given that the food industry suffers from a lack of credibility, food producers have elected to forge close relationships with government and non-government health bodies and with the scientific health community more broadly. Health benefits are being value-added through nutrition messages and logos, and are an important ingredient in the fight for competitive advantage. This chapter shows that chicken producers are not the only ones to use health and nutrition as a source of endorsement. Importantly, appeals to medicine offer retail traders a chance of attaining some form of authority status in relation to food: whether charismatic, rational-legal, or traditional (Weber 1947). Fashioning patterns of authority between retailers and consumers is a significant process in the fortunes of the chicken meat commodity complex; a matter that is further detailed in the next chapter.

The marketing and selling of health benefits means servicing consumers in much the same manner as servicing producers with product development ideas.[9] Like the producer services sector, a consumer services sector has evolved which consists of government agencies, private firms and professional groups who provide education, skills, information and risk assessment tools to assist people to consume particular products and services and not others. Consumer service providers may be enlisted by retailers and producers for their particular insights of the cultural milieu of consumers, or they may operate in the market place of their own volition, pitching their offerings directly to the consuming public. Most televised food programs, recipe books and food and wine festivals exemplify more autonomous cultural goods production. Whereas producer services provide economic capabilities to corporations, consumer service providers supply cultural capabilities to both corporations and consumers and, in the words of Steven Lukes (1974), they facilitate the 'mobilisation of bias' through shaping desires and beliefs.

Holding the distinction between producer and consumer services becomes difficult when the same agency, firm or professional group simultaneously services producers and consumers. Sassen acknowledges this issue and identifies what she calls the 'mixed business consumer-producer service'. The producer-consumer service, a tidier term, provides a critical interface between producers and consumers by providing logistical support for retail operations and household consumption operations. This support typically includes copywriters and advertisers communicating information about the store and loyalty programs, nutritionists working on communications with consumers regarding healthy meals, special store event organisers and the in-store product promoters.

The provision of services both upstream and downstream by supermarkets is becoming commonplace, with supermarkets providing product growing inputs like technical back-up to producers (for example, video-imaging of fat marbling in meat), while providing product information to consumers (for example, designating meat fat content on products). In this sense supermarkets are more than retailers and even processed food producers: they broker relationships between producers and consumers on their terms. Supermarkets use the producer-consumer services sector to sell lifestyle.

Some retail traders are moving beyond selling health benefits and are presently engaged in a risky cultural adventure to decouple the connection between home and meal preparation. Indeed there is a certain urgency to redefine what constitutes the home given that the food market is characterised by finite demand. Intervening in household production and consumption is a hazardous endeavour because commercial firms cannot be seen to undermine significant cultural institutions like the family. It is not accidental then that they are promoting minimal home-meal preparation as a service that will allow the former family cook 'freedom' to pursue paid work while serving a 'proper meal'. The present move to have meals prepared outside the home and variously consumed within or outside it requires the construction of new understandings about the home and gender roles, family relations and of entire culinary cultures that are based upon the mother as family cook. A good way to describe this is the notion that food retailers are configuring themselves 'as a quasi-substitute parent, a sort of paternalistic mothering figure, guaranteeing the consumer's comfort and security, but simultaneously setting [themselves] apart from their lived experience' (Fiddes 1995, p. 144). Reconfiguring corporate identity, as we saw with KFC's move away from an image of fried food purveyor to one of family food provider, is a substantial undertaking: one that is only possible because of the parasitic nature of the interests of actors in the spheres of distribution and exchange.

CONCLUSION

Chicken meat producers admit that ten years ago they controlled the balance of power in the commodity complex but that now it is 'the turn' of the supermarkets. This chapter highlights the importance of cultural production activities for capital production where chicken is concerned and argues that the balance of power has shifted in the direction of those retailers and allied cultural producers who are involved in trading practices, ideas and values. Producer-consumer services are responsible for the capital-in-circulation acquiring new or enhanced values and they are pre-eminent in establishing the boundaries in which the shaping and negotiation of cultural practices takes place. The research outlined in this chapter supports the point that 'demand is a socially regulated and generated impulse, not an artefact of individual whims or needs' (Appadurai 1986, p. 32). It confirms that lifestyles can be exchanged and corporate images refashioned through the trade in commodities. The family-friendly supermarket and fast food chain are offering to take more of the dual income so that any household member, from microwave-adept age onwards, can buy time and nutrition through minimal home-based food preparation. They are restructuring which groups are the feeders in society.

This conclusion justifies more attention being paid to cultural production activities and to the circuit of reproduction. Chapter 8 reflects upon the proposition that understanding power in the culinary culture and commodity complexes means appreciating the ways in which cultural and economic processes interact.

8
REASSEMBLING THE CHICKEN: A CULTURAL ECONOMY VIEW OF POWER

Power, like energy, must be regarded as continually passing from any one of its forms into any other, and it should be the business of social science to seek the laws of such transformation.

(Russell, cited in Bourdieu 1977, p. 235)

The use and application of power frequently enters into changes in a society's food consumption habits. Where this power originates; how it is applied and to what ends; and in what manner people undertake to deal with it, are all part of what happens when food habits change. We do not understand these processes as well, even though they are of immense importance to the world's future.

(Mintz 1996, pp. 17–18)

This chapter proceeds to take up the challenge posed by Russell and Mintz, whose statements about power open the chapter. It focuses on a theme that permeates across Chapters 4 to 7: that capital flows are aided and circumscribed by cultural processes as surely as they are by political economy processes, such as government regulation, or by the physical features of the landscape and the natural characteristics of the commodity. While it is relatively straightforward to identify the distribution of power in a commodity complex, it is quite another matter to nominate the manner in which power is produced, reproduced and changes form. The social life of the chicken helps us, however, to understand more clearly the nature of power in complex industrial societies. The transformed life of this pre-modern bird tells us, for example, that the exercise of power in the early 21st century can only be understood through the interdependence of cultural and economic processes. It suggests that personal experience of the material world through contact with knowledge, technologies and ideas is as significant as material production. Furthermore, by identifying the value-adding that is attached to products as a cultural rather than an economic phenomenon, the activities of production, consumption, distribution and exchange can be recast as truly social and not simply economic activities. The product design process, retailing practices, food knowledge and discourse production highlight how even the most economically powerful actors must negotiate daily with the cultural landscape. This engagement suggests that Polanyi (1944) was premature to worry that culture had been marginalised by political and economic institutions. He was correct, however, to note the relative autonomy between cultural, political and economic activities.

The purpose here is to sketch what is entailed in adopting a cultural economy view of power. Cultural economic activity has been described by various sources as the exercise of symbolic power (Bourdieu 1984); the mobilisation of reflexive accumulation processes by a new middle class linked to a trade in words and symbols (Lash & Urry 1994; Zukin 1991); and as the interaction between market and household or moral economies (Silverstone et al. 1992). Some academics are suggesting that a circuit of culture is operating in much the same way as the circuit of capital, outlined in Chapter 2. I describe below the contours of a cultural circuit framework and argue that a cultural economy perspective energises the modified commodity analysis framework used in this book to explain the dynamics of commodity complexes and culinary cultures. Before embarking on the conceptual task, I summarise the social life of an Australian table chicken and what that life tells us about the distribution and production of power in a commodity complex and in the broader culinary culture.

THE SOCIAL LIFE OF THE CHICKEN

It is worth reiterating that just forty years ago chicken meat was obtained primarily outside the capitalist market place. In a short period, the backyard chook shed and associated exchanges inspired by barter, *luck*,[1] and gift-giving have been almost totally displaced by market place exchanges. A once-prized food is still prized but for different reasons; the result of company manoeuvres rather than a variety of household activities. Nevertheless, the social life of the chicken appears to be a particularly rich life if one considers how it is produced in 2002. It begins its existence in the scientist's laboratory, where all the skills of a craftsman are brought to bear on genetic structures. It then moves from one factory farm to another. The multiplication farm and then the hatchery are where the scientists' progeny are tended by unskilled but highly personalised labour. The hatchery is a high-tech factory, with imported skills embedded in the machinery and computers. Indeed, the hatchery could well compete with hospital nurseries for computerisation. The chick then moves onto the factory growing farm to be tended by the often income-rich landed labour, working under contract to millionaire businessmen and women or, from time to time, to public shareholders. On this farm, one multi-skilled person, with the help of unskilled labour, makes use of sophisticated machinery which they own to tend tens of thousands of living creatures. Once the chicken is six weeks old it is ready for a whirlwind engagement with Fordist and possibly post-Fordist systems. In the space of twenty-four hours, the chicken will move through up to four different forms of production: the hard-to-categorise factory farm; the Fordist processing factory; the post-Fordist supply chain; and the retailer, who may or may not use craft methods to create differentiated products.[2] The diverse forms of production and distribution, over a short period, guarantees the mass-produced chicken a special status: not as a rare food but as the height of efficiency. Chicken production is a model aspired to by other agrifood producers.

The mass-produced chicken and negligible alternative denies the presence of a third food regime, or a regime driven by consumers. Still, mass production does not preclude a cultural dimension to commodity status and success. In each decade since the sixties, chicken's low price has been accompanied by potent messages and material practices. In the 1960s, it was the widespread availability of a festive food, while a decade later it was freshness that resulted from naturalising the product through technological developments. (Ironically, mass 'naturalisation' was a response to the demands of environment and counter-culture movements). In the 1980s the focus was on nutritionalisation and the demonisation of fat, producing a backlash against red meat. In the 1990s the predominant value was the availability of convenient products to suit the lifestyles of over-committed family cooks. Importantly, the earlier meanings attached to chicken have not been displaced by the more recent ones. Rather, a layering of values has taken place. Chicken, it seems, can withstand innumerable meanings even if they contradict one another. Indeed, the term 'halfway food' used by one consumer is an appropriate metaphor for chicken's contemporary status. Chicken is halfway between healthy and unhealthy; indulgent and frugal; effort and ease; a traditional home-cooked meal prepared by a stranger; a special food that is now everyday. Distribution-centred networks have positioned chicken meat as ideal for delivering a relevant regime of values for pressured times: convenience, price, health, family harmony and mobility. Chicken delivers such a melange of values with less effort than other meats.

DISTRIBUTING POWER: THE BALANCE OF POWER WITHIN THE CHICKEN MEAT COMMODITY COMPLEX

The literature reviews in Chapters 2 and 3 provide opposing views of powerful actors in the food system. In the main, agrifood analysts and retail geographers emphasise those responsible for capital flows as being the sole driving force behind food commodity complexes. Furthermore, Chapter 2 outlined the Marxist framework for understanding the life of commodities in these terms. That particular framework posits different forms of capital — money, productive activity and commodities — circulating across the different circuits of production, realisation and reproduction. It has been used to good effect to explain capital's spatial dynamics: the way in which capitalist relations extend across time and space, and transform commodity complexes both globally and nationally (Pritchard 1995). This is summarised in Table 8.1.

Table 8.1
Circuits of capital

CIRCUITS OF CAPITAL	MOMENTS OF CAPITAL
Production	P – C production capital to commodity capital
Realisation	C – M commodity capital to money
Reproduction	M – P money to production capital

However, the table chicken story shows this framework to be inadequate to account for all the processes that constitute a commodity complex. In addition, the Marxist inspired regulatory regimes schema of Fordism/post-Fordism is too crude to grapple with primary production, in particular. The 'agrarian question' persists precisely because of the cultural and natural dimensions to agricultural production. What is needed is an approach for embedding capital in a wider set of social relations and activities.

It is in this context that the focus groups offer considerable insight. At one level they support much of what is emerging from the sociology of consumption, especially consumer questioning of experts and criticisms of the market. Such behaviour has been termed 'reflexivity' (Beck et al. 1994; Lash & Urry 1994). While these particular consumers, like a majority of Australian consumers, are not creating alternative commodity production and distribution systems, their questions and complaints act to constrain the power of producers and distributors in the market economy.[3] In addition, their sometimes factually incorrect fears over hormones and birdcages, for example, act as a brake to producer power by promoting a perception of ethically flawed producers.

Overall, however, I would argue that the type of consumer reflexivity that is being practised in relation to food confers limited power only. In many instances, reflecting on the food supply leads to defeatism and inaction. The consumers who were interviewed for this research took for granted food availability, but were troubled by the values attached to individual foods. In particular, they were concerned with a food's contribution to nutritional and family harmony values. Their anxiety about the overall food supply led them to desire improved information about some food products, to believe myths about others and to demonstrate ignorance about the remainder. They were eager to know which foods would contribute to their personal well being and less enthusiastic for information about the production process, particularly where animals were concerned. They wanted a part of the story only and, ironically, their desire to remain ignorant of the production process contributed to misplaced anxiety about the food supply. Furthermore, the consumption of chicken by my respondents suggests that standards surrounding what is good to eat are not that high. The consumers were prepared to make significant concessions to chicken: around its healthiness, safety and lack of distinctive taste. They exhibited a marked propensity to say that they appreciate chicken's low fat status on the one hand, and to consume fried, sauced and fatty roast chicken on the other. They admit to concerns for animal welfare and yet overlook the stress brought upon the birds by intensive farming. They rank food hygiene as one of their greatest concerns yet ignore chicken meat's proneness for microbial contamination. Overall, chicken's ability to incorporate a host of contradictions, especially a desire for healthy convenience, adequately summarised consumer sentiment. Despite their concerns, chicken offered a degree of relaxation not afforded by other meats and, given all the other pressures in their lives, the consumers were not disposed to agitate for improvements to its quality.

The proposition that consumers *en masse* are powerful was supported by the widespread practise of thrift through the pursuit of cheap foods, but it was undermined by the extent to which consumption appears to be contingent on information and practices produced outside consumers' social groupings. The influence of media coverage of expert opinion and nutritional claims made by particular health authorities is extensive. This point is made more significant when considering that children are assuming a status as social actors in the food system through impacting upon household food decisions. Their use, noted in Chapter 7, of nutritional claims to obtain fast food is a case in point. In some households, children have become as potent in food decision-making as husbands.

These explanations of the norms of consumption surrounding chicken indicate varying degrees of personal power, authority, coercion, influence and control, to use the distinctions provided by McIntosh and Zey (1989). However, this personal power was exercised as individual shopper, exerciser of thrift, manager of the body and convenor of family life. Personal resources were not systematically invested in collective attempts to challenge producers. In the case of the chicken meat complex and other fresh food complexes, it is the traders of goods and symbols who have emerged as the challengers to producer power by placing themselves at the intersection of cultural and economic activity. On the Australian landscape for more than three-quarters of a century, not only are Woolworths and Coles the largest merchandisers of fresh food, they expend enormous resources creating commodity contexts which resonate with the idea of supporting simultaneously the quality producer and the knowledgeable, caring consumer. Unlike the producers, who are out of sight and hence out of mind, the traders are both familiar and familial.

The present day cultural activities of corporations would not surprise sociologists of trade. For example, Evers and Schrader (1994) note that over the centuries, traders have engaged in transferring wealth into symbolic capital in order to strive for respectability. Friedberg (1997) highlights the necessity of traders to mobilise non-economic attributes as a way of fostering commodity exchange. Norms of trust and reputability based around identities of the trader are commonplace in analyses of pre-capitalist societies, but Friedberg (1997) argues the applicability of these same social processes for understanding dynamics in industrial societies. She analyses how those involved in bringing food from the farm to the kitchen deal with questions of risk and trust and she describes two situations to reveal the building of 'trust capital'. In the pre-capitalist market a trustworthy reputation is achieved by direct reputation, but in the capitalist market of the 21st century it is attained through indirect association with reputable others. Her argument resonates with my own about Australian supermarkets and producers forming alliances with medical authorities and other professional groupings in order to themselves become *good to think*. In short, a vital process in a hazardous venture like producing and selling food is the symbolic regulation of markets through corporate reputation.

The pre-eminence of food traders has arisen precisely at the same time that a mother's cooking is less pivotal to her value in the family. As she becomes subject to the authority of the capitalist labour force and her paid work activity becomes more important to labour markets and household economies, the mother's traditional food authority diminishes. Thus, as the actor most central to household food provisioning moves to the periphery of that activity, a greater fluidity is introduced across social arrangements more broadly. An interesting question remains: how is this fluidity occurring?

REPRODUCING POWER: THE EMERGENCE OF MARKET-BASED AUTHORITIES

The table chicken story indicates that relations of power between producers, retailers and consumers are produced, transformed and reproduced through a series of processes including the capacity to accumulate capital, consumer experience of material practices, the symbolic regulation of markets and the authorisation of market-based players. Only the first of these elements concerns the exercise of economic power while the remaining three involve the exercise of different facets of cultural power: beliefs and practices, symbols and expertise. These four semi-autonomous processes contribute to what has been called 'reflexive accumulation', a form of growth less dependent on the sale of goods than the trade in services, communications and information to generate wealth (Lash & Urry 1994, p. 64). Reflexive accumulation strategies provide a basis for the amassment of cultural resources — status, reputation and ideas — by producers, distributors and consumers, which in turn may be traded for market influence.

The capacity to accumulate large amounts of economic capital is a feature of both the production and distribution sides of the chicken commodity complex. In the Australian setting, both sides exhibit high levels of industry concentration but attract differing levels of government regulation. The heavy regulation of production assists the capital accumulation of the producers (Dixon & Burgess 1998) just as the absence of regulation fosters rapid capital turnovers and huge profits by the retail chains (Chapter 6). How is it though, that in the last decade the supermarkets have assumed the balance of power *vis a vis* the producers? The answer lies in the way that the large retail traders are able to influence the food provisioning practices of households. Their capacity for influence stems from their superior knowledge of consumers as well as their ability to communicate a regime of values through their product ranges. The mobilisation of these particular resources sees retail corporations contributing to the process by which social groupings and social status form; a contest in which 'the power to impose the legitimate mode of thought and expression ... is increasingly waged in the field of the production of symbolic goods' (Bourdieu 1977, p. 170). And, as we saw earlier, even something as mundane as chicken meat operates as a potent symbol of a particular lifestyle.

Bourdieu not only highlights the existence of other forms of capital beside finance capital, including symbolic and cultural capital, he points out that all such forms could be accumulated, exchanged and transformed. The table chicken story confirms that these other forms of capital are targets of control and struggle, just as is finance or productive capital. Nowhere is this better illustrated than in Chapter 7, with the tale of KFC's rebirth as a family food provider. Despite the millions of dollars spent by PepsiCo, the world's largest food retailer, consumers could not be persuaded to buy a whole roast chicken from a fried-snack purveyor. Consequently we saw the corporation reinvest retail capital to reconfigure its image as more family friendly, and they did this through purporting to sell meals not snacks.

The on-going campaign being waged by KFC, as well as the supermarkets, over where foods should be cooked — in the household kitchen, the so-called community kitchens of fast food outlets or the industrial kitchens supplying supermarkets — is but a recent manifestation of the types of negotiations that traders need to enter into with households. The basis for contemporary negotiations is laid through the psychographic research undertaken by corporations. Coles Myer uses The Mind Map to negotiate its way through the consumer life-world. In the mid-1990s, this market technology revealed that the mass of Australian consumers still hold to 'family values' and 'conservatism', or a desire for stability as opposed to change. Despite a century of advertisements promoting the value of constant change and the perpetual search for meaning through consumption, household belief systems continue to be relatively autonomous of marketised belief systems.

Material in Chapter 2 mentions the long-standing tenacity of women's food provisioning for family life and Chapter 4 illustrates the interference by household moral economies to unfettered market power. The distinction between moral and market economies is useful for the way it highlights problems in the transformation of meanings attached to commodities as the latter move from the market into the household. In work that emphasises the shift between public and private spheres, the household as a moral economy is:

> ... conceived as part of a transactional system of economic and social relations within the more objective economy and society of the public sphere. Within this framework households are seen as being actively engaged with the products and meanings of this formal, commodity and individual-based economy. This engagement involves the appropriation of these commodities into domestic culture — they are domesticated — and through that appropriation they are incorporated and redefined in different terms, in accordance with the household's own values and interests (Silverstone et al. 1992, p. 16).

It is conceivable that E.P. Thompson's moral economy of outward-looking claims against rulers and local authorities for basic rights and goods has given way to Miller's moral economy of demands on the market to support particular familial and domestic arrangements. This shift arguably reflects the activities of those who are trading in information, services and care work. For as Bourdieu (1984) notes, today's market economies rely upon 'new professionals' who are working in a 'substitution' industry where words are exchanged rather than goods. The primary role of these professionals, who are increasingly being employed by corporations, is to mobilise reputations and bias in order to shape regimes of value. The importance of their symbolic production activity lies in the idea that has occupied sociologists for a century or more: that power relations are a struggle over what is thinkable.

Thus far I have been addressing power as the capacity to act: as having superior access to finance capital and to cultural resources. I have argued that freedom from government regulation of their economic power, coupled with their capacity for symbolic production, allows the large food traders to frame the terms of debates about food systems, food practices and family life. But as sociologists of power point out, capacity to act is but one dimension of the exercise of power; the other precondition is the right to act — to be seen as a legitimate exerciser of power (Hindess 1996; Lukes 1974).

The key to the legitimate exercise of power, in my opinion, lies in the concept of authority status, and I am arguing that consumers have authorised market-based players, particularly food retailers, as legitimate exercisers of power. For centuries, authorities have been based in civil society or the state, with authority status contingent on the possession of charisma, technical resources (including skills and expertise) or traditional rights to govern (Weber 1947). Sennett reminds us, however, that the source of authority can change because authority 'is itself inherently an act of imagination — it is a search for solidity and security in the strength of others' (Sennett 1980, p. 197). Sennett argues that authority is an emotional bond between people who are unequal and, as such, the exercise of authority has political consequences.

As traditional authority in respect of culinary cultures is challenged or becomes more ephemeral, people look to a range of specialist fields for replacement authorities. It is here that food consumers turn to those in the market with food-related expertise. What appears to be happening, at least in the case of supermarkets and image chains such as KFC, is that they are assuming *de facto* authority while the *de jure* authority of the family cook is on the wane.[4]

In an era where there are few determinate authorities and many claimants to authority, the quest to be perceived as legitimate is an onerous one (Giddens 1991, pp. 194–96). For this reason, food retailers are co-opting the services of health professionals and are associating themselves with medical authorities. Through these third party associations market-based actors themselves become more authoritative. If as Sennett (1980, p. 165) alleges, the work of authority is 'to convert power into images of strength', then the actors striving to control the activities of those both upstream and downstream in the food system are well placed symbolically to do so. Thanks to the efforts of countless advertising and marketing professionals, a charisma of strength is communicated at every opportunity: shareholder meetings, store promotions, nightly television advertising, and prominent siting at the heart of suburbs and shopping precincts. At the same time as corporations are claiming strength, they communicate that previous culinary orders are *passé*. Domesticity is currently portrayed in KFC advertising as a mother who exhibits knowledgeable concern, not necessarily cooking skills, and while the family is still shown eating together, this is in an array of settings, not necessarily the home kitchen or dining room. The media portrayals both reflect and confirm how the mothers who participated in this study see themselves: they are no longer wholly responsible for feeding the family but have become convenors; assembling family members and organising meals, eating times and places.

The numerous networks that comprise the food producer-consumer services sector, many of which are anchored around supermarket and fast food chains, have assumed responsibility for managing food-related risks, both real and perceived, and for promoting regimes of value to suit the interests of network members. Neither activity, however, is straightforward because the values attached to a commodity or a culinary culture result from a host of transactions conducted over decades by a host of actors, including consumers as bearers of family and culinary traditions, and as producers of myths and tastes. In addition, the extraordinary array of actors implicated in *making culinary culture* do not agree among themselves about the meanings being promoted, transmitted and practised. Conflicts are particularly manifest around the standards used to judge commodity status. Chicken-based contestations, for example, are apparent in relation to debates over chicken's contribution to personal health and the place of convenience foods in the diet. Just as the state has been managing conflicts between chicken growers and processors, so various social experts, some state-sponsored, some commercially-sponsored and others self-appointed, adjudicate contestations surrounding the status of the commodity. In the Australian context at least, the influence of government food regulatory authorities is diminishing and at this point in time, the authority relation is tipped toward the market-based players and away from consumers.

CULTURAL AND ECONOMIC INTERPENETRATION

On the basis of what has been learnt about the social life of chicken, it is not too contrived to argue that the esteem with which chicken is held is built upon its movement between spheres of cultural and economic activity. To paraphrase Bourdieu (1984), the operations of national commodity complexes are the by-product of struggles over economic and cultural power, especially symbolic power. Others have argued similarly about the symbiotic relationship between cultural and economic processes: that '[n]ot only is the cultural construction of meaning and symbols inherently a matter of political and economic interests but the reverse also holds — the concerns of political economy are inherently conflicts over meanings and symbols' (Marcus & Fisher cited in Gregory & Altman 1989, p. 37).

Despite mounting evidence of the interpenetration of cultural and economic processes, the relationship between the sphere of culture and capital accumulation remains ill-defined in sociology. In one attempt to bring them together, Zukin identifies a circuit of cultural capital as 'a key to understanding the structural linkage between cultural and economic values today' (Zukin 1991, p. 260). Furthermore, she questions whether 'the continuous production of cultural commodities, moving between 'economic' and 'cultural' circuits, continually increases the economic value of investment capital' (Zukin 1991, p. 260). However, it remains unclear whether she was referring to the flow of cultural goods or the flow of capital that results from cultural production activity. Other social scientists have played with the notion of a circuit of culture. David Harvey, for instance, urges analysts of capital circulation to consider whether cultural production lies within the domain of the circulation of capital or not (Harvey 1996, p. 67). Again, he does not resolve the matter but he does nominate a number of 'critical' cultural 'moments' in what he terms the 'social process', besides material practices and social relations. These are discourse/language, power, beliefs/values/desires and institutions/rituals (Harvey 1996, p. 78).

My research into the social life of the chicken indicates the operation of a set of cultural circuits and moments, which support the conjecture of both Zukin and Harvey that a circuit of culture may exist. The cultural circuits that I am suggesting here stand in close relationship to the circuits of capital outlined in Table 8.1, but are relatively autonomous. On the basis of the table chicken story, I am proposing three circuits of culture. The first circuit is commercially inspired and involves cultural production activity being subsumed into the circuits of capital. Examples include the corporate repositioning of Woolworths as fresh food provider, the evolution of Coles' All Things Poultry stores, and the promotion of KFC as family food provider. In each instance, labour was devoted to a range of cultural economy activities: product development and differentiation strategies, product packaging, advertising, and other promotions. I believe that this type of activity is what Zukin had in mind for her 'circuit of cultural capital'. The second cultural circuit, I propose, is socially inspired: moral and household economies are incorporated here. The moral economy shows how culture can be situated within material and social realities without becoming derivative of these realities (Thompson 1993, p. 13). Social cultural circuit practices that reflect this principle include observance of traditions, folkways and religious activities. The third circuit is emotion-based. While emotional activity is relevant to economic activity it is in no way beholden to it. Beasley (1994) identifies how goods and services can be produced on the basis of love, affection and care, or what she terms an emotional economy. Orthodox economists might call the activity in this circuit 'irrational' because of inefficient use of resources, and reliance on feelings rather than 'facts' to guide decision-making.

Within the three circuits are specific moments. The commercial-cultural circuit involves the relatively non-contentious movement between wants and needs, or use values. The use values are shaped further in the circuits of capital as they become exchange values (generally the price paid and opportunities foregone in appropriating a good or service). In many writings, the commercial-cultural circuit is subsumed into the circuit of capital reproduction. The social-cultural circuit involves the struggle over control of household, family and community life: including the values that should underpin the interface between labour markets and social relationships. In one of his refinements to the concept of habitus, Bourdieu highlights that patterns of behaviour must be conceived as 'the virtue made of necessity', and as 'a product of the incorporation of objective necessity' (Bourdieu 1990, p. 11). In terms of the social relationships and labour market interface, some social scientists are pointing to the emergence of flexible dispositions, moral codes and familial and social ties as a direct result of flexible labour markets (Elchardus 1994). While flexible dispositions suit the labour market requirements of many industrial nations, labour market flexibility has been credited with encouraging consumers to be unmanageable and opportunistic (Gabriel & Lang 1995, p. 190). Finally, the emotional-cultural circuit involves the movement between valuing the past and the present, stability and change, the new and the old, the young and the aged without necessary regard for their economic value. This is where omnivore's paradox is found, where anxieties and hopes reside and where authority status is negotiated. The dynamic in this circuit is best encapsulated as the struggle between commitment and flux over beliefs, relationships and practices.[5] Table 8.2 summarises what I consider to constitute the circuits of culture.

Table 8.2
Circuits of culture

Circuits of culture	Moments of culture
Commercial	Translating wants into needs
Social	Reconciling virtue and necessity; family life and labour markets
Emotional	Acting out the tensions between commitment and flux, desire and anxiety; entering into authority relations.

The circuits of social and emotional culture are relatively autonomous from the circuits of capital because alternatives, albeit limited, to the market exist, and resistance to market dynamics persists in some quarters. Household production, reciprocity and redistribution may be in abeyance, but if the new corporate providers of family foods cannot continue to deliver cheap chickens and to curb the risks of their products then the backyard as food source could re-emerge, and a parent as family cook could continue into the foreseeable future.

As a commodity moves between the circuits of capital and culture it acquires economic and cultural values. It is for this reason that those driven to accumulate capital must play a role in shaping desire, alleviating anxieties and managing moral and household economies. One of the challenges facing the producer-consumer services sector is to keep the chicken commodity moving so that it roosts in the commercially inspired circuit. It is not surprising that large retailers acting as traders of goods and social practices, in conjunction with other producer-consumer services, are exercising the balance of power in many fresh food commodity complexes: it is they who are orchestrating the social life of these commodities. This situation is likely to remain for as long as the producer-consumer services continue to influence three areas: the investment of capital in symbolic power creation, the locus of authority status in respect of the food system, and the moral economy demands which are placed on the food system.

If a political economy focus is the exercise of economic power in the capitalist economy, a cultural economy perspective emphasises the power that is implicated in cultural activity. Building on the work of Polanyi, Halperin argues that 'the term "cultural economies" refers to an analytic perspective which examines economies as they are embedded in and constructed by cultural systems that are more powerful than particular individuals and particular historical moments' (Halperin 1994, p. 17). In other words, a cultural economy perspective should be conceived as a distinctive ontological approach for understanding social arrangements because it privileges neither cultural nor economic processes.

At the heart of any given cultural economy are two components: valuation processes; and the power that different actors derive from, and exercise in, their position in these processes. Chapter 3 describes both the pivotal role of wage relations to production, and the centrality of commodity relations to consumption. Despite renewed interest in distribution and exchange, the defining relation of these spheres of activity remains unresolved. The social life of the chicken reveals the patterning of authority relations to be highly significant as an enabler of, and constraint to, social action. The quest for authority status to help consumers and citizens make sense of the world is critical to those in commodity production and distribution. In a work that theorises the centrality for social action of a collectivity's sense of coherence the point was made that people do not need to feel personally in control, but that 'the location of power is where it is legitimately supposed to be' (Antonovsky 1979, p. 128). Consequently, it is feasible to argue that the authority relation, or 'the legitimate exercise of domination' as defined by Weber (1947), is the determining relation in distribution and exchange. Acquiring a reputation as a legitimate source of wisdom, expertise, knowledge, goods or service is arguably the major imperative of many who are engaged in distribution and exchange.

In summary, I believe that by overlaying the circuits of culture on the circuits of capital a framework emerges that adds a process-like dimension otherwise missing in frameworks inspired by political economy approaches. A twin circuits commodity framework, represented by bringing together Tables 8.1 and 8.2, restores dynamism by suggesting a degree of contingent fluidity in forms of power. It makes transparent the multidimensional nature of cultural production and symbolic power, and the importance of various processes of valuation to capital accumulation. Resistance to economically driven meanings is highlighted in the emotional cultural circuit, while processes of negotiation are apparent in the social cultural circuit. The twin circuits approach makes it possible to ascertain how particular actors or actor networks assume the balance of power from time to time. The twin circuits represent what I call a cultural economy approach for studying commodity complexes and power more generally. The approach is summarised in Figure 8.1. I have borrowed the idea for the schematic representation from Harvey's depiction of the key moments in the social process (Harvey 1996, p. 78).

This depiction of power is a far cry from that given in Table 3.1. A cultural economy perspective goes beyond the wage relation and the commodification process. By acknowledging the practices, beliefs and discourses of the consumer, producer and trader, a cultural economy framework acknowledges both the fraught processes of valuation and the importance of 'emotional' activity in shaping power relations. In this way a cultural economy perspective does not simply permit a description of the distribution of power in commodity complexes, but can explain shifts in the balance or reproduction of power.

Figure 8.1
Moments in the social life of a commodity

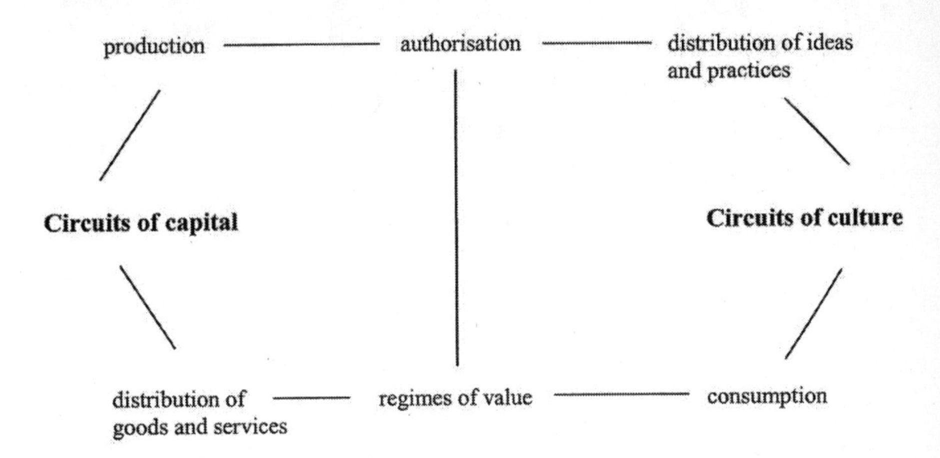

CONCLUSION

The Australian table chicken story confirms that popular foods continue to depend on more than multi-site, low-price availability. These two economic values are enjoined by a simultaneous circulation of cultural associations: images of healthy bodies, portrayals of healthy social relationships and healthy corporate activity, and ideas about the responsible management of the food system. Indeed, my research suggests that these cultural attributes are more important than price in a food supply characterised by relative abundance and cheap foods. Chicken, with its ability to effortlessly cross boundaries, lends itself perfectly to a cultural economy based on forms of food production and distribution which have marginalised women's household labour and magnified the economic significance of women's food service sector labour. The final chapter examines whether this scenario has a global dimension.

9 THE GLOBAL CHICKEN

Australians are among the highest consumers of chicken in the world, but they are not alone in their fondness for chicken. The worldwide consumption of poultry almost overtook beef consumption in 2001;[1] a milestone Australia will reach by 2013 (Instate 1997). The new global esteem for chicken suggests the presence of a chicken meat production regime that extends far and wide, and Boyd and Watts (1997, p. 194) claim that a global poultry industry does indeed exist. Not all are impressed with chicken's global credentials, however. In the opinion of Bonanno and colleagues, 'there is not as yet, in fact, a global chicken. Rather, chickens as a commodity, while beginning to become internationalized, are still more of a regional market intended for localized markets' (Bonanno et al. 1994, p. 10). The second part of their assessment is correct because most countries are self-sufficient in chicken meat: only eleven per cent of poultry enters world trade. Since the Second World War, local production for domestic purposes has been the international norm with few exceptions. Put another way, 'poultry industries worldwide are multi-domestic rather than multinational' (Larkin & Heilbron 1997, p. 3).

While a focus on the economic dimensions of chicken meat production provides few features that normally indicate globalisation, I nevertheless propose that there are grounds for talking about a global chicken. This chapter describes the multifarious routes to chicken having a global presence and the consequences of that presence. The cultural economy approach which is adopted in this chapter reveals how cultural inputs — technologies, knowledge and material practices — are being exchanged, thereby facilitating both the worldwide production and consumption of chicken meat and a global culinary dynamism based on healthy convenience.

For Appadurai, global economies are the result of five cultural economy flows: people, technology, global capital, electronic dissemination of information, and ideas and state ideologies (Appadurai 1990, p. 6). The circulation and exchange of these ethno-scapes, techno-scapes, finance-scapes, media-scapes, and ideo-scapes spread resources, values and ideas widely and together pave the way for globalisation. Waters (1995) has added international tourism to Appadurai's list. The global chicken story shows how these flows coalesce in culinary cultures.

What follows here is a description of the ways in which globalisation is manifest in the production, distribution and consumption of chicken. By comparing these facets of the social life of chicken meat with the trajectories of other global foods, I explore the claim that a global chicken is making an appearance. Furthermore, by contextualising international chicken meat consumption within broader social and economic trends, it becomes clear that chicken is itself contributing to processes of economic and cultural globalisation. Chicken meat commodity complexes not only exemplify the operation of the twin circuits of capital and culture described in Chapter 8, they play a part in extending the global reach of particular culinary cultures. As a result they contribute to reproducing relations of power centred on the reflexive accumulation of capital.

PRODUCTION REGIMES

According to *Tucker in Australia*, the animals which landed with the First Fleet of British colonisers in 1788 included 'one stallion, three mares, three colts, four cows, twenty-nine sheep, nineteen goats, forty-nine hogs, twenty-five pigs, five rabbits, eighteen turkeys, twenty-nine geese, thirty-five ducks, one hundred and thirty-two fowls and eighty-seven chickens', and 'only pigs and poultry did well' (Wood 1977, p. 33). Obviously the poultry was reared for their eggs, because while pork and some beef appeared in the rations of the early settlers chicken was not mentioned.

This long-distance exchange would no longer take place, unless the animals were destined for an Australian zoo. In order to protect existing stocks from disease, government quarantine regulations prohibit most animals from entering the country, in a living state at least. In addition, because chickens are relatively easy to rear in all but the coldest climates, they do not need to be traded across the seas. And in the case of Australia at least, the success of local bird breeders has been sufficient, until recently, for producing chickens for a mass market. Again Australia is not unique.

Chicken meat production throughout the world has grown in all regions except Eastern Europe (Tomoda 1997).[2] Production increases have been most impressive in the developing countries of China, India and Brazil. Given the multi-domestic aspect of this production, the significance to the industry of global production methods, transnational firm activity and international supply chains deserves examination. In relation to Australia, foreign investment in the industry is negligible and, with the exception of Kentucky Fried Chicken, transnational corporations headquartered outside Australia have little direct industry influence. Nevertheless, the current wave of global economic activity is beginning to be felt by chicken farmers and processors as successive governments embrace free trade policies and labour markets are benchmarked to be competitive with those of Australia's major trading partners (see Dixon & Burgess 1998; Fairbrother 2001).

While chicken meat is not traded widely between countries, there is an extensive north-to-south trade in genetic avian stock and the requisite protein meal and medicines. As an Australian Benchmarking Study commented: '[c]ountries can easily achieve high degrees of domestic self-sufficiency through the ready availability of internationally traded production equipment, technology and feed while protecting their domestic markets through high tariff protection and complex veterinary and health restrictions' (Larkin & Heilbron 1997, p. 23). Many nations, including Australia, depend on European vaccines as well as European and American incubation and processing equipment.

Underpinning the trade in these inputs is the widespread adoption of a technology for co-ordinating production and supply, namely vertical integration and contract farming.[3] Figure 5.1 summarised the workings of vertical integration relevant to the Australian industry, but that figure can be replicated for many countries, as can the power relations that result. Just as some commentators attribute the success of poultry industries to their adoption of vertical integration, others have decried the accompanying contract farming. The influence of vertically integrated poultry processors over farming is alleged to have increased processes of subsumption, thereby diminishing the power of family farmers (see discussion in Fulton & Clark 1996). Subsumption is a process by which capital penetrates production processes, and the term 'indirect subsumption' is used to describe the situation in which the agricultural labour process becomes beholden to technological inputs supplied by others (Whatmore et al. 1987). In the case of poultry production, the relevant technological inputs are the genetic stock, veterinary advice, farm machinery and contracts for selling farm products to processors and retailers.

While the majority of chicken meat production is the result of vertical integration and indirect subsumption, the latter is an uneven process that must be understood historically and spatially. Caution is warranted when one analyses the power of Australia's chicken farmers relative to poultry farmers in other countries, who work under vertical integration regimes. As a result of investing large sums of their own money in building dedicated broiler growing farms (Cain 1996), Australian chicken farmers exert considerable influence over the agribusinesses for whom they labour. Moreover, their aggregate commercial strength is supported by government legislation overseeing the basis on which they receive incomes for growing birds. Australia's chicken farmers believe that their better remuneration, in comparison to the United States at least, can be attributed to government regulation that allows them to negotiate collectively with processors. Following a study tour of broiler operations in America and Britain, one delegate concurred that, in the southern states of America where most broiler production occurs, 'the contract format often is weighted in favour of the processor', and 'considerable grower mistrust of the major processors has been engendered, and not without foundation'. He concluded that 'the manner in which absolute power has been exercised in some instances in the US regardless of legal or moral considerations is viewed by the author as offensive. US growers feel they are being consciously denied security of tenure (arguably fundamental to man's well being) as a bargaining ploy' (Taylor 1995). This is the sort of argument that is used by farmers in Australia to resist the international benchmarking of labour costs.

Australian farmers, however, do not fear comparisons with farmers in developed countries as much as they fear imports of cheap chicken from developing countries. They are worried by an initiative of the World Bank's International Finance Corporation (IFC) which is financing poultry projects in developing countries. The IFC has commented that '[p]oultry production is an efficient way to produce affordable animal protein, which is increasingly in demand in low and middle income countries. The poultry trade also has major benefits to small farmers who are associated with the processing industry' (IFC 1995, p. 8). While this approach encourages self-sufficiency in the final product and is a far cry from a food-aid mentality,[4] it nevertheless plugs local producers into a network of transnational avian stock and feed producers, and pharmaceutical and equipment manufacturers located in the Northern Hemisphere.

In the near future, the poultry industry worldwide is likely to witness more traditional forms of trade in final product. In the Asian region, agreements made between Asian countries under the ASEAN Free Trade Area (AFTA) and the World Trade Organisation (WTO) come into effect in 2003 and 2005 respectively. Under AFTA, import tariffs are to be reduced to less that five per cent by 2003, and it is anticipated that the well advanced poultry industry in Thailand and the fledgling Chinese industry will expand their exports to the Philippines, Indonesia, Malaysia, Japan and Taiwan. Thailand, in

2000, also had a substantial export trade with the United Kingdom, Germany and the Netherlands: a trend that is expected to consolidate with the BSE crisis in Europe and the Foot and Mouth Disease outbreak in the United Kingdom over the last two years (Rabobank International 2001a). Despite such opportunities, specialist poultry development agencies such as the Dutch Rabobank continue to caution that trade liberalisation is only the most obvious barrier to the global trade in poultry. Equally significant to the free movement of chicken products are the cultural barriers of religion, for example where the meat is required to be Halal (Indonesia), or of a certain sanitary standard (Australia), or a particular colour (white in the United States and dark pink in Japan).

DISTRIBUTION

Until recently, the most significant feature of the global distribution of chicken meat has been on the input side. However, American companies which continue to over-produce chicken meat have been targeting the Pacific region as an export market (Aull-Hyde et al. 1994; Boyd & Watts 1997). The United States is by far the largest chicken meat exporter. In 1997 the Australian government agreed for the first time to accept imported chicken meat. The decision to allow imports from the United States and Denmark occurred after considerable public debate and came with a major caveat: the meat was to be cooked at such high temperatures that it would be fit for pet food only. In essence, this was a continuation of the ban on chicken meat imports (Dixon & Burgess 1998). Since then, the industry has continued to lobby the government to prevent chicken meat imports, emphasising the bio-security threats to the local bird flock by exotic strains of avian diseases (Fairbrother 2001). Thus far, the tightly organised and highly concentrated Australian chicken meat industry has been able to exploit the benefits of international trade in inputs and to turn their resulting market strength to fight further inroads by international interests.

As highlighted in Chapters 6 and 7 however, there is more to distribution than the movement of goods. In the case of chicken, supermarkets and the global image chains have been responsible for exchanging ideas about values and practices associated with buying and eating chicken. While more research is needed on the extent of global retailing, evidence exists that Western retail principles are becoming widely adopted (Shackleton 1996) and some market analysts have attributed chicken's success to highly visible and successful fast food outlets, in particular KFC (Rabobank 1993). This particular company commenced its international trajectory in the 1960s, and by 1996 had more than 10 000 stores across ninety-four countries. Even with concessions to local tastes, each of these stores carries the same imagery and messages about American-style capitalism, chicken and food more generally.

In a recent analysis of the poultry market in the United Kingdom, Rabobank International (2001b) explained that country's strong growth in per capita consumption by two factors: retailer led positioning of poultry as 'the principle meat range' and a shift from whole, frozen birds towards higher value-added and convenience products. The bank laid particular stress on the 'strong power of the food retailer', describing how consolidation among the biggest food retailers has meant that six retailers are together responsible for seventy-five per cent of chicken sales. Like Australian supermarkets, retailers in the United Kingdom are being more demanding of their suppliers, favouring a small number of innovative processing firms, capable of supporting the retail customer. To date, the supermarkets have relied on British firms but it appears that the cost advantages enjoyed by Brazilian and Thai firms may see them supply more processed meat products in the future (Rabobank International 2001a).

The significance of the activities of retailers in shaping culinary cultures has been noted by Appadurai (1986, p. 33). Unlike ruling elites who act as custodians of established tastes and restricted exchange, merchants and traders have long promulgated new tastes. Interestingly, one of Australia's elite agricultural groups — beef producers — can be readily identified as protecting a well-established culinary culture based on a main meal of red meat and three vegetables. Their long association with family-owned butchers, who exchange familiarity, tradition and specialisation, is part of that culture (Australian Supermarkets Institute/AC Nielsen 1998). In contrast, Australia's chicken producers exhibit little opposition to the supermarket trader's desire to foster new tastes: a fact that might explain why chicken farmers have attracted such odium within Australian agriculture.

CONSUMPTION

The Dutch Rabobank concluded its global assessment of the poultry industry by commenting that '[w]orldwide demand for poultry products has increased substantially in both developed and developing countries at the expense of beef and pork consumption. The price, value, and religious acceptability have been favourable for demand' (Rabobank 1993, p. 21). The IFC (1995) and other sources more recently show an increasing poultry consumption trend for most countries (Rabobank 2001a; United States Department of Agriculture Foreign Agriculture Service 2002). Growth between the mid-1990s and the turn of the century has been most pronounced in Asian countries, where annual consumption varies enormously between Hong Kong at fifty-seven kilograms per person (the highest in the world) to less than one kilogram per person in India. The only countries to show negative growth are Russia, some Eastern European countries such as Romania, and Kuwait.

While the favourable price of chicken is repeatedly offered as a reason for its success, the World Bank's IFC (1995) and the Rabobank (1993; 2001a; 2001b) agree that chicken meat consumption can no longer be fully explained by the dynamics of prices and income. Increasingly, 'cultural' factors are being added to explain chicken's popularity, with a typical assessment attributing 'dietary health, variety, and convenience [as] characteristics of poultry products that have increased demand' (Rabobank 1993, p. 21; Rabobank 2001b). These are the same reasons offered by the experts quoted in Chapter 4 regarding Australia's fondness for chicken. Here, I amplify these and other commonly mentioned cultural attributes in the international context.

Substantial support exists for the argument that health concerns have reinforced chicken's low price to make it a popular choice. Both Levenstein (1993) and Mintz (1996) accord nutrition science great importance in their respective histories of culinary cultures, and Williams (1997) observes that an appreciation by high SES groups of the links between good health and good nutrition has been especially potent in chicken's trajectory for that particular group since the 1960s. The rise of what is known as the *body culture* has coincided almost exactly with fears about cholesterol and the often erroneous views about the role of red meat in causing high cholesterol levels. Indeed, nutrition science perhaps is the biggest contributor to the falls in meat consumption over the last two decades in the meat-eating settler states of the United States, Canada and Australia.

In addition to health attributes is the idea of convenience, which should not be viewed simply as a property of availability, but by ease of preparation and eating. Chicken has become desirable in European-based societies because civilising in the form of table manners does not have to take place: it is quite acceptable to eschew knife and fork when eating chicken wings and drumsticks (Visser 1986). Indeed, KFC has made 'finger lick'n good' a promotional highpoint for nearly thirty years.[5] As processors now recognise, certain chicken portions make the perfect finger food and as such chicken is ideally suited for the eating practice identified as grazing (Wray 1995). Furthermore, chicken has always had a place in cultures used to street food. Asian migrants, whose culinary cultures generally favour chicken and pork (Instate 1997), have represented these cultures in many industrialised countries since the 1970s.

Adding to its cross-cultural credentials, chicken is one of the few traditional foods that has not attracted religious, economic or social taboos. The lack of specific taboos around its killing and consumption means that it is easier than other meats to assimilate into the diet because few groups find it objectionable. Chicken reveals the importance of secular religions too, with the 1960s social movements of the counter-culture and personal growth creating a favourable commodity context. Moreover, as pre-secular observations of the distinction between everyday and festive foods have dissipated (Falk 1991), the table chicken straddles with ease the categories of 'ordinary' and 'special'. The idea that ordinary people should consume special chicken has been present for four centuries. King Henri IV is reputed to have said in the 16th century: 'I hope to make France so prosperous that every peasant will have chicken in his pot on Sundays', while in the 1920s the United States Republican Party used the slogan 'A chicken in every pot' in their election campaign (Visser 1986; Whit 1995). Its cross-culinary status is reflected further by the fact that chicken-based meals can successfully incorporate a basic contradiction: along with fish, chicken is acceptable to some vegetarians (Beardsworth & Keil 1992). Like fish, it can be dismissed as *non-meat* and *non-animal.* Furthermore, its perceived lightness as a food possibly contributes to a sense that one is doing less harm to oneself and to the environment by eating chicken.

Finally, Mintz has made important observations about the connection between the whiteness of foods, purity and esteem. He argues that the whiteness, and its associations with purity, explains the success of sugar, rice, flour, milk and chicken breast (Mintz 1996, p. 89). Australian consumption data shows that colour could indeed be associated with the relative esteem of meats: crimson offal, red beef, pink lamb and mutton, pale pink pork and white chicken can be placed on a continuum from large to slight falls in consumption between the mid-1980s and mid-1990s (Dixon 2000). However, it remains unclear whether colour is the relevant cause of this, given the widespread campaigns to lower dietary fat intake and the higher increase in retail price of redder meats over paler meats during this time.[6] Instead, chicken may stand for 'eating virtue' in a way that other meats cannot, more in keeping with Mintz's theory of purity.[7]

While there is no consensus about the relative importance of the attributes leading to chicken's esteem, the International Finance Corporation (IFC) has statistics for the early 1990s that show a certain inevitability about meat consumption. Each of the major meats — beef, pork and poultry — have similar relative demand trajectories in developed and developing countries, and between rural and urban populations within developing countries. The IFC notes that the demand for meat increases with rises in per capita income and urbanisation, and International Labour Organisation (ILO) figures support the case that as incomes rise in relation to the cost of living, consumers spend more on 'protein products of animal origin' (Tomado 1997). Furthermore, the IFC figures indicate that chicken and pork enter as the preferred meats for urban dwellers earning wages. However, when incomes equate with those of the United States and Northern Europe two trends occur: the traditionally 'big' meat eating countries such as Australia and America show a decline in overall meat consumption, with chicken meat's decline being the smallest; and beef consumption starts to rise among wealthier groups. This is the situation in Japan. In Australia, earlier declines in beef consumption have slowed as beef is being eaten in moderation by higher socio-economic groups once again and among women who have responded to the campaigns that they should consume foods rich in iron (Shoebridge 1995). The available data supports a proposition that changing preferences for particular meats are in line with the changing nature of economic and urban development. On the basis of the preceding material, it is possible to speculate that the industrial process in the West at least has been fuelled over the last century by specific meats, irrespective of national culinary cultures. 'Tell me your preferred meat and I will tell you where you are on the development index' is worthy of further research.

THE MAKING OF GLOBAL FOODS FROM 1880 TO 1970: LABOUR MARKET REQUIREMENTS, GEO-POLITICAL MANOEUVRING AND SOCIAL EMULATION

The histories of three earlier global commodities — sugar, beef and wheat — help us to understand more about the factors which contribute to the interrelationship between culinary dynamism and globalisation. The histories reveal both continuities and change in respect of chicken.

In *Sweetness and Power*, Mintz (1985) details sugar's rise to prominence in Britain and Europe between 1650 and 1900, and its subsequent appeal in the United States. In later work he reflects that he had not succeeded in locating a single cause for sugar's consumption (Mintz 1996, p. 18). Instead, Mintz shows that in the space of two and a half centuries, a mix of political, economic and social factors were operating within a context of the expansion of the world economy to make some foods and types of meal more or less acceptable. When explaining sugar's trajectory as a working class food he returns to a mix of class rivalry in symbolic construction, to imperial policies which led it to be cheap and available, and the need for a particular calorific intake by the working class. Mintz stresses the relationship between what he describes as 'the world-market solution for drug foods', such as sucrose, and the spread and consolidation of industrialisation. He maintains that:

> [s]ubstances like tea, sugar, rum, and tobacco were used by working people in accord with the tempos of working class life ... Sugar was taken up just as work schedules were quickening, as the movement from the countryside to city was accelerating, and as the factory system was taking shape and spreading. Such changes more and more affected the patterning of eating habits (Mintz 1985, p. 174).

Moreover, sugar's 'consumption was also a symbolic demonstration that the system that produced it was successful' (Mintz 1985, p. 174). As the system of work changed, foods that suited this system had to be found to feed the new labouring classes.

At the end of the 19th century two new wage foods were added to the working class diet in industrialising nations: red meat and wheat. Friedmann and McMichael's (1989) explanation of how the trade in these commodities underpinned wider capital accumulation and regulatory regimes is generally accepted. According to Friedmann (1990; 1994), the livestock complex evolved through systems of integration between feed producers, feedlot technology and intensive livestock producers. She describes how being able to differentiate between animal feed grains and human food grains created grain surpluses leading to American-produced wheat becoming a wage food of third world proletarians after World War 2: '[t]he historically privileged grain of Europe, the mark of wealth and status, became the wage food of twentieth century proletarianization' (Friedmann 1990, p. 201). Instead of the 19th century periphery-to-centre flows of foods such as sugar and oils, wheat was produced at the centre and shipped as food-aid to developing countries. The actions of states and farm sectors to dispose of grain surpluses underpinned what was referred to earlier as the second food regime.

The history of the industrialised world's esteem for red meats is brief compared to that of sugar. In an exceedingly short time, meat was displacing complex carbohydrates and vegetable based proteins. In 'A short meat oriented history of the world', Cockburn describes the lack of variation in the European diet until the middle of the 18th century:

> Grains took up about 90% of a family's food ... From the moment that the victuallers and provisioners of the Napoleonic wars pioneered the organization of the mass production line and also modern methods of food preservation, the stage was set for the annihilation of both time and space in matters of food consumption. The vast cattle herds that began to graze the pastures of the western US, Australia and Argentina signalled the change (Cockburn 1996, p. 24).

Science, in the form of animal genetics and husbandry, and modern transport systems, were essential to red meat production and availability at a cost affordable to working people. Others have similarly proposed that the transnationalisation of bovine productive processes has met the needs of working people for foods which symbolise physical strength, muscle power and restoration (Fiddes 1991).

Ruling class foods, such as sugar, red meat and wheat, have been adopted by working classes not simply as a result of a social game of emulation, to paraphrase Bourdieu (1984), but because geo-political manoeuvring has led key social actors to make them available. In the space of a century two quite distinctive food regimes have supplied the needs of proletarians while fuelling figuratively and economically the geo-political formations which surrounded the extensive and intensive phases of capital accumulation.

Government regulation of national economies, the labour market requirements of manufacturers and the activities of multinational firms have been potent shapers of the second food regime. A key feature of the Fordist labour market has been its attraction to women who want their families to participate fully in what many refer to as 'consumer society'.[8] Using British data, Goodman and Redclift (1991) reveal the extent of the interdependence of women working for wages outside the home and the demand for industrially produced food. They assert that changes in food consumption since the 1950s are linked 'to the shift from the production of use values in the home, to the production of exchange values outside the home' (Goodman & Redclift 1991, p. 28). Supermarkets have been especially responsive, targeting women in the labour market with services of convenience, culminating more recently with offerings of home-meal-replacement.

THE MAKING OF GLOBAL FOODS FROM 1970 ONWARDS: THE STANDARDISATION OF SOCIAL AND ECONOMIC ASPIRATIONS WITHIN FLEXIBLE LABOUR MARKETS

The adoption of sugar, red meat and wheat as wage foods occurred at times of relative scarcity of food choice as well as relative undernourishment of working people. In contexts where food choice is abundant, one has to look anew at how more traditional foods in both industrialised and industrialising nations are being displaced so rapidly by chicken meat: what many agree to be a relatively tasteless food (Fiddes 1995; Symons 1982; Visser 1986; Whit 1995).[9] Despite its myriad forms and numerous symbolic associations, chicken products are the result of remarkable standardisation in production and distribution systems and, as Sanderson (1986) has argued, internationalisation proceeds through the 'internalization' of international industry standards.

Sanderson arrived at this conclusion by analysing what he calls 'the world steer'. He found that beef products involve the international standardisation of producer technology and social relations along lines that firstly are transnational in scope (for example, American feedlot technology, European antibiotics and Japanese markets for boxed beef), and secondly approach an international standard for consumption and trade (for example, immunity from major contagious diseases and certain meat characteristics). Sanderson's model does not require transnational corporation involvement for the presence of globalisation, but rather internationally adopted technologies and organisational forms. Following Sanderson's line of thinking, being a multidomestic industry does not preclude the building of a global chicken. Instead, a commodity that results from the local, standardised, assembly of parts supplied from many parts of the world is sufficient. This process has been likened to the contemporary automobile industry, minus the corporate badging. If one compares the production and consumption of chicken meat in many countries with that of the 'world steer', its global status is confirmed.

Sanderson implies that there exists a global consumer, ready to buy the internationally acceptable beefcut. Yet agreement over this existence is not widespread. For instance, Jussaume cautions that 'it may be presumptuous to extrapolate ... that similarities in working conditions on shop floors around the world are fostering convergence in all aspects of daily life' (Jussaume 1991, p. 49). Nevertheless, whilst employment trends, including the mix of casual, full-time and part-time work, and average weekly hours of paid work, are not uniform across industrialised countries, what is emerging is that most OECD countries are embracing women's entry into the paid workforce. Furthermore, what commenced in the 1960s as 'flexible working hours', expressed through flexitime arrangements (Elchardus 1991), has come to mean the demise of the standard working day, week and year, and also the end of high wage levels across the working class in a large number of OECD countries (Burgess et al. 1996, p. 10). In these countries at least, given the interdependence of labour routines and household agreements, a context is created in which foods are re-evaluated. The commodification of household cleaning, cooking and gardening are both a reflection of households coping with labour force changes and a part of the ongoing marketisation of everyday life that began in the 19th century. These processes extend to very ordinary activities such as the ability to cook from scratch (Bittman & Pixley 1997; Counihan 1988; Gofton 1995).

Whereas Sanderson's emphasis is upon economically driven standardisation within production, it is important to be alert to the process of cultural standardisation. As shown in Chapter 7, the cultural standards, or regimes of value, surrounding home life, meals and family life are being rewritten with particular consequences for women. In a study of how McDonald's intervenes in the relationship between the family, food and citizenship, Probyn provides a startlingly similar assessment:

> In the intense mobile of food, family and the citizen what we have is a continual displacement: a shifting game of substitution, condensation and reversal. Perhaps most striking of all is the way in which this site of intensity no longer appears to need women. As it produces a motherless familial citizen, McDonald's plays an essential role in the space cleared by the dispersal of women ... In part, this is a game of substitution: the previous triangle of food, femininity and the family needs to be rearranged in order for McDonald's to take its place in the discourse of the family, food and citizenship (Probyn 1998, p. 169).

While there may not yet be a global consumer, significant resources are being expended by transnational corporations to influence how citizens of the world think about their relationship to the food supply. These are the new cultural standard bearers who anchor the producer-consumer services food sector: a sector of professionals who accept being geographically and occupationally mobile and cosmopolitan in outlook (Sassen 1991; Zukin 1991). Indeed, it is tempting to suggest that any emergent global food standards will be underpinned by a particular community of production-consumption for which there is recent sociological agreement: the middle class consumer group, which has high disposable income derived from working as professionals in the service industries (Harvey 1989; Lash 1990; Sassen 1991). This group has been credited with the evolution of nouvelle cuisine in the United States (Zukin 1991) and the emergence of ideo-cuisines in that same country (Sokolov 1991). In Australia as elsewhere, the major supermarket chains are chasing this group who are expected to embrace the specialist poultry stores and the relatively expensive supermarket home-meal-replacement goods and services. Equally, these producer-consumers are family members who share the struggles of large numbers of family cooks in their pursuit of contemporary moral economy values such as the right to 'good' food and the primacy of children's needs.

In a context of competing demands, a familiar principle is operating to guide food choices: that '[t]o be accepted, new ideas about food must also fit in with people's social and economic aspirations' (Levenstein 1988, p. 211). And at the beginning of the 21st century, the aspirations that are in the ascendant are oriented around participation in the global market place. Producer-consumer services are central to the incremental shifts in aspirations because they legitimately mediate the interface between the labour force and households. Through their involvement in the discursive practices concerning meals, homelife and good food they are reinventing themselves as food authorities and they are shaping the mouth of the community to accept what they themselves desire and have to offer.

Whereas Mintz argues that sugar fuelled the British Industrial Revolution through its centrality in working class diets, and Friedmann proposed that wheat has been fundamental to the diets of working people across the world for much of the 20th century, I surmise that chicken is becoming the preferred protein and meat of working people in numerous countries. While the colour of chicken may be differently hued across the globe depending on its feed, its breed and the processing methods used, chicken contains the features of foods which are demanded by a low wage and urbanising labour force, as well as by a high wage, health conscious labour force. As earlier sections testify, these features are an undeniable polyglot but they converge around the practice of healthy convenience.

CONCLUSION

The social life of the chicken shows that esteemed foods no longer need to belong to rulers or to be traded across the seas. Rather what is traded are manufacturing systems, production system standards, retailing strategies and internationally recognised symbols. In addition, food tastes travel via immigrants and tourism as well as the cultural standards of the new food authorities, with each wave enlarging culinary cultural imaginations. The mass media's coverage of the latest innovations from the nutrition sciences is equally potent in shaping tastes. At the same time, social systems including labour markets and organisational practices travel via the activities of transnational corporations and global agencies. Together these cultural economy flows interact with the social and the emotional worlds of consumers to make culinary cultures dynamic. Given the ability of ideas, symbols and people to travel relatively effortlessly, it is not surprising that contemporary culinary cultures resonate with flexibility synonyms: grazing, multiculturalism and ideo-cuisines.

In addition, the table chicken reinforces the importance of a single food commodity to the smooth operations of capitalist economies, even when producers have less influence over commodity systems than they have had in the recent past. This is because commodities can be used to signify the robustness of political and economic systems. Chicken, it seems, has performed this function over the centuries, and it is presently being used to communicate that the global economy is one of healthy progress. The cultural economy of the chicken highlights the manner in which power relations are being reproduced at this point in history: through more local forms of authority being usurped by market mediated authority, with today's experts being located in the global marketplace. Furthermore, claims that consumers are responsible for food systems change and culinary cultural dynamism are inadequate. Instead, the social life of the chicken reveals that while socially constructed tastes are powerful, individual consumers are not.

NOTES

1 BARBECUES, CHICKEN SHEDS AND CULINARY DYNAMISM

1 In a review of the progress of the field of food sociology, culinary culture was defined as 'a shorthand term for the ensemble of attitudes and tastes people bring to cooking and eating' (S Mennell, A Murcott, & A van Otterloo, *The Sociology of Food: Eating, Diet and Culture*, Sage, London, 1992, p. 20). For my purposes the food system encompasses practices associated with food growing, preparation, exchange and consumption. The culinary culture and food system are interdependent.

2 The agrarian question refers to the uncertainty around whether those making a living from the land have escaped capitalist social relations. 'The ambiguous class location of the rural petty bourgeoisie (who, as small capitalists own and control production but, who as workers, use their own labour power in the creation of value) gives rise to ideological confusion' (G Lawrence, *Capitalism and the Countryside: The Rural Crisis in Australia*, Pluto Press, Sydney, 1987, p. 109).

3 Mintz's scheme is built upon data showing that in a wide variety of cuisines there was a core meal pattern comprised of starchy foods, or complex carbohydrates found in grasses and rhizomes; legumes or pulses which are calorie and protein carrying plants to complement the starchy plants; and a fringe accompaniment of flavours, found in relishes etc. Mintz argues that '[a] few hundred years ago, the ancient and widespread core-fringe-legume pattern, highly diverse in detail but ... surprisingly similar from case to case in its organization and broad outlines, began to crumble' (S Mintz, Eating and being: what food means. In B Harriss-White (ed.) *Food: Multidisciplinary Perspectives,* Basil Blackwell, Cambridge, 1994, p. 111).

4 'Agro-food' is a term used in the United States, the equivalent term being 'agrifood' in Australia and New Zealand.

5 Giddens' theory of structuration distinguishes between structures, conceived as providing frameworks for action, and the implementation, or practising, of the rules and resources offered by institutions. The process of structuration has a dual character: it both enables and constrains action and is changed or reinforced by action (D Held & J Thompson, *Social Theory of Modern Societies: Anthony Giddens and his Critics.* Cambridge University Press, Cambridge 1989, p. 4).

2 POWER IN THE CULINARY CULTURE

1 The seminar was sponsored by RMIT's Transport Research Centre, Melbourne on 15 May 1997.

2 A commodity complex has been defined 'as a chain (or web) of production and consumption relations, linking farmers and farm workers to consuming individuals, households, and communities. Within each web are private and state institutions that buy, sell, provide inputs, process, transport, distribute, and finance each link. Each complex includes many class, gender, and cultural relations, within a specific (changing) international division of labor' (H Friedmann, Distance and durability: shaky foundations of the world food economy. In P McMichael (ed.) *The Global Restructuring of Agro-Food Systems,* Cornell University Press, Ithaca, 1994, p. 258).

3 The term Fordism was coined by Gramsci in the 1930s to refer to the systematic changes occurring in capitalist societies resulting from car manufacturer Henry Ford's introduction of mechanised assembly line methods for the mass production of consumer goods.

4 From a brief description of family farms in Australia in the early 20th century one gathers that mixed farms offered women similar opportunities and heavy work loads to grow, sell and barter food, particularly eggs and vegetables (D Cahn, Australians in the early twentieth century. In B Wood (ed.) *Tucker in Australia.* Hill of Content, Melbourne, 1977).

5 'Just In Time production methods refer to strategies based on inventory control, keeping inventories to a minimum and having parts delivered "just in time" for their assembly' (J Mathews, *Tools of Change: New Technology and the Democratisation of Work,* Pluto Press, Sydney 1989, pp. 79–80).

3 CONSTRUCTING THE SOCIAL LIFE OF THE CHICKEN

1 Wallerstein's world systems theory (1980) is acknowledged by H Friedmann and P McMichael (Agriculture and the state system, *Sociologia Ruralis,* XXXIX (2): 93–117, 1989) as influencing their food regimes theory.

2 W Friedland (Reprise on commodity systems methodology, *International Journal of Sociology of Agriculture and Food,* 9(1): 82–103, 2001) has since added three 'research foci' to the 1984 model, including 'sectoral organization' to include state sector regulation of the commodity. The other foci nominated includes the 'commodity culture' which refers to cultural forms found either among commodity producers or consumers, and 'scale', referring to the spatial and social relationships underpinning the commodity.

3 H Friedmann (Family wheat farms and third world diets: a paradoxical relationship between unwaged and waged labour, in J Collins & M Gimenz (eds), *Work Without Wages,* State University Press of New York, 1990, pp. 193–213) has argued that the second food regime is characterised by a close fit between the regime of production and norms of consumption and that agrifood producers in league with governments engineer the symbiosis.

4 A delocalised food supply has been defined as one where 'an increasing proportion of the daily diet comes from distant places usually through commercial channels' (G Pelto & P Pelto Diet and delocalization: dietary changes since 1750, in R Rotberg & T Rabb (eds), *Hunger and History: The Impact of Changing Food Production and Consumption Patterns on Society,* Cambridge University Press, Cambridge, 1983, p. 309).

5 A Giddens (*The Consequences of Modernity,* Stanford University Press, Stanford, 1990, p. 21) uses the term 'disembedding processes' to refer to knowing something without being directly involved or understanding what is happening.

6 Some of Australia's largest food manufacturers, such as Goodman Fielder,

have adopted corporate nutrition policies which refer to the latest nutrition science research findings.

7 Across the United States, the American Dietetic Association (ADA) has convened conferences with titles such as 'Supermarkets: Nutrition Learning Centers of the Future'. One such event was described thus: '[s]upermarkets have become centers for delivering foods and nutrition information to customers. More and more Americans use the supermarket as a resource center for information on the nutritional value of foods and for more healthful ways to prepare these foods. The key to influencing the consumer's food selection in the supermarket is to identify the most effective method of providing shoppers with nutrition information designed to change behavior' (American Dietetic Association, brochure, 1990).

8 Supermarkets can not only claim superior insight into consumers, their knowledge of health matters and concerns far outstrips that of producers. In a British study of the food chain's knowledge of the latest thinking regarding health matters and dietary changes, food retailers 'were the most aware and had reacted the most ... with farmers being found to know very little about these matters' (J Darrell, The response of the food chain to healthy eating, *British Food Journal* 94(4), pp. 7–11).

4 CONSUMING CHICKEN: BUYING TIME, NUTRITION AND FAMILY HARMONY

1 Stephanie Alexander was chef at Stephanie's Restaurant, one of Melbourne's fine dining restaurants for most of the 1990s.

2 The basis for classifying the groups on SES lines is spelt out in Dixon (*Cooks, chooks and culinary cultures: a cultural economy model for the study of food commodity systems,* PhD thesis, Royal Melbourne Institute of Technology, 2000).

3 A full description of this aspect of the research can be found in Dixon (*Cooks, Chooks and Culinary Culture*, 2000).

4 Australians eat very little poultry other than chicken. This contrasts with the United States where turkey is popular, and with Asia and some European countries where duck is consumed in significant amounts.

5 'Most people perceive shopping for meals as hard work — not as a relaxation. (Other sociological research suggests that only 10-12% see it as a relaxing and pleasurable experience)' (Steggles, The Fresh Chicken Market: Trends & Dynamics, Steggles Internal Document, 1996, p. 29).

6 Citizenship right was the fourth relationship or 'process of provisioning' (A Warde, Notes on the relationship between production and consumption, in R Burrows & C Marsh (eds), *Consumption and Class: Divisions and Change,* Macmillan, London, 1992, p. 19). It applies mostly to services and goods in which the state is a major actor. State provided food applies particularly to times of war and shortage, and to food aid programs. None of the participants in the focus groups mentioned this form of provisioning.

7 Two households had a hen house and explained that this was for two reasons: to obtain free-range eggs and to acquaint children with more traditional foodways.

8 Interview with Safeway Delicatessen manager at Moonee Ponds

9 This is consistent with the finding that the market is used by those in the workforce to reduce time spent in unpaid work activities, including purchasing more pre-processed goods (M Bittman & S Mathur, Can people buy their way out of housework? Paper presented at the eighth World Congress of Sociology, July, Bieldfeld, Germany, 1994, p. 7).

10 I observed that the adults ate most of the nuggets while the children played with them.

11 Australia's third largest supermarket chain at the time.

12 Symons noted that during the 1960s, the 'biggest sales of frozen foods were

achieved by poultry' (M Symons, *One Continuous Picnic: A History of Eating in Australia,* Duck Press, Adelaide, 1982, p. 209).

13 Australia's most high profile and arguably its most respected animal welfare organisation.

14 For Thompson, rulers took it upon themselves to respond to mass unrest, especially food riots, on the basis that their subjects should not starve. The food riots were an example of subjects exercising their rights to particular goods and they represented an acceptance by governing bodies that they had responsibilities in relation to 'the immorality of ... profiteering upon the necessities of the people' (E P Thompson, *Customs In Common,* Penguin Books, London, 1993, p. 337).

5 PRODUCING CHICKEN: WORKING WITH REAL TIME

1 The terms chicken, poultry and broiler are used interchangeably in Australia and the term broiler is used almost exclusively in the United States. While I will refer to those who grow chickens mainly as chicken farmers, I will use chicken grower when that is how others refer to them.

2 Bartter acquired Goodman Fielder's fifty percent holding in Steggles Foods Products Pty Ltd, a joint venture with American food group, OSI International Foods.

3 The term 'New Australians' refers to non-English speaking migrants.

4 One aspect of intensive livestock production which is particularly criticised by environment groups is the use of grain to feed animals as opposed to humans. As Eco-Consumer put it, '[i]n developed countries the animal feed industry is the largest consumer of grains. The poultry industry worldwide accounts for one-third of the total production of feed, with grain making up 70% of poultry broiler feed ... The unsuitable agricultural practices used to produce those grains rely on high energy inputs, synthetic chemicals and pesticides' (Eco-Consumer, Chicken welfare: still an issue, *Eco-Consumer Newsletter* February, 1996 p. 1–3). However, as another student of the industry has pointed out, '[w]e should bear in mind that poultry farming is the most efficient way of converting grain into flesh; other meat industries consume much more agricultural produce' (M Visser, *Much Depends Upon Dinner,* Penguin Books, London, 1986, p. 141).

5 In 1997 mortality rates of broiler flocks averaged about five per cent in Victoria, with rates of two to three per cent being considered desirable by the industry.

6 This claim was contested by two processors, including the one who initially made the claim but it was attested to by several process workers.

7 The Senate Committee on Animal Welfare heard of smaller plants where the birds are not stunned before decapitation, and it took evidence of the time and degree of suffering experienced. It recommended that stunning prior to slaughter was the most humane practice.

8 A small Victorian processor promotes its chicken as 'processed chemically free'. It begins by saying: 'Traditionally, chicken is processed by being placed in a bath of chemicals and water. By this method the flesh is bleached and the natural juices that are an integral part of 'real chicken' are leached out' (La Iionica Poultry n.d.). Some processors I spoke to do not believe it is possible to defeather without the scalding process.

9 The booklet, *Manual Handling and Noise in the Poultry Processing Industry* recommends such practices (Occupational Health and Safety Authority, *Manual Handling and Noise in the Poultry Processing Industry,* Department of Business and Employment, Victoria, n.d.).

10 In a domestic household a stainless steel kitchen knife would last a lifetime, whereas industry knives are sharpened several times daily and thrown out after six to eight weeks.

11 Processors want to have a two shift operation, or to operate up to seventeen hours, a day. In Victoria at least, only one shift is allowed mainly because plants are sited close to residential areas.
12 Especially salamis and sausages.
13 Although the interviewee also noted that 'no matter what you do to beef, it's red, it will always look heavier than a white piece of chicken'.

6 'HERE A CHOOK, THERE A CHOOK, EVERYWHERE A CHOOK CHOOK'

1 A line from the nursery rhyme 'Old McDonald Had A Farm', being sung by a small girl as the tram passed by a McDonald's store at the Royal Children's Hospital.
2 The material which is presented in this chapter comes from six main sources: supermarket company documents which are in the public domain; interviews with supermarket and poultry shop personnel; site visits to several supermarkets, two poultry shops and the headquarters of the two biggest supermarket chains; various market analysis reports; attendance at a national poultry industry convention; and data compiled by Coles Myer Research Division.
3 The specific gains to consumers, identified by the Committee included: deregulated trading hours, greater product choice, lower food prices and the convenience of one-stop shopping.
4 These two choice meat cuts are priced in a manner which makes other parts of the pig and chicken almost worthless to the producers.
5 In January 1997 Safeway Moonee Ponds was selling breast fillet at $7 per kilogram, close to what they bought it for from the processor, but skinned breast fillets were retailing for $9.50; this is where the profit was being made on that day.
6 It seems that history is being repeated. Kim Humphery's history of Australian supermarkets contains a photo from 1963 of a Coles' New World supermarket which shows an earlier space rocket perched on top of the roof of the Dandenong store (K Humphery, *Shelf Life: Supermarkets and the Changing Culture of Consumption,* Cambridge University Press, Cambridge, 1998, p. 101).
7 'Other retailers report that fresh food generates thirty to thirty-five per cent of their sales, compared with less than twenty-five per cent five years ago' (N Shoebridge, Fresh food boom demands quick action, *Business Review Weekly,* 18 July, 1994, p. 40).
8 Consumers' behaviours, beliefs, fears and hopes are under continuous scrutiny. Large retailers have extraordinary quantities of data pertaining to the demographics and psychology of Australia's population. For example, at the heart of the Coles Myer operation in the mid-1990s was market research that segmented the population on the basis of individuals' expressed optimism–pessimism, income levels and spending disposition, and degree of individual-social orientation. The Mind Map was the trade-marked name of the particular psy expertise to which it subscribed. As a result, distinctive communities of consumption have been identified and on this basis decisions are made in three areas: the siting of stores, micro-marketing strategies and store product ranges.
9 Coles' preferred suppliers include the two largest processors plus a medium sized processor in each state.
10 'Efforts to cut costs in a highly competitive industry mean increasing attention to more cost effective grocery warehousing and distribution. Cross-docking is being hailed as at least part of the solution' (Anonymous, Around the table: cross-docking, *Supermarket,* June, 1995, p. 9).
11 Furthermore, it may provide a smokescreen to suggestions of anti-competitive behaviour.
12 Labour process accounts of fast food chain operations are becoming numerous and the practices of chains, such as KFC, have been described elsewhere. Suffice

to say, fast food chains rely on Fordist production and distribution processes. This is summed up well by PepsiCo, the owner of KFC, which boasted that the company has been so profitable because 'Our people ... want to be the finest. We hire eagles and train them to fly in formation' (PepsiCo Operations Fact Sheet n.d.).

13 Ryan and Burgess (The Supermarket Co. In J Burgess, P Keogh, D Macdonald, G Morgan, G Strachan & S Ryan (eds) *Enterprise Bargaining in Three Female Dominated Workplaces in the Hunter: Processes, Participation and Outcomes.* Employment Studies Centre Working Paper Series No. 26, University of Newcastle, 1996) found shop floor workers to be very dissatisfied with their union, which they compared unfavourably to the butchers' union.

14 In supplementary remarks to the Committee, Senator Murray of the Australian Democrats argued that 'there is a clear link between the dominance of the majors, and the extent of trading hours deregulation' (Joint Select Committee on the Retailing Sector, *Fair Market or Market Failure: A Review of Australia's Retailing* Sector, Parliament of Australia, Canberra, 1999, p. 137).

15 Women are also highly segmented within the workforce, with two-thirds of women workers falling into three of the twelve Australian Bureau of Statistics industry sectors. These are community services, wholesale and retail trade and the finance, property and business service sectors' (B Van Gramberg, Women, industrial relations and public policy, in L Hancock (ed.), *Women, Public Policy and the State,* Macmillan Education Australia, South Yarra, 1999, p. 104).

7 DISCURSIVE PRACTICES OF THE CHICKEN

1 In 1998 it was not uncommon to travel on a tram which goes by the market, with women of different nationalities sitting with boxes at their feet, and the boxes containing one or two live chickens.

2 The only evidence for this is the development and sale, but not unqualified success, of non-fried chicken products.

3 Shortly after the appearance of the insert, the Australian Meat and Livestock Corporation (AMLC), which represents the combined red meat industries, sponsored a sixty page supplement, 'The nutrient composition of Australian meats and poultry', in the journal *Food Australia.* APIA, representing Inghams and Steggles, took the AMLC to the Trade Practices Commission charging the red meat body with selectively using data from the *New Idea* insert to misrepresent chicken in its advertisements (J Fairbrother, The poultry industry: technology's child two decades on, *Food Australia,* November, 1988, p. 462).

4 This point is supported by an interview with supermarket executives regarding the 'fresh food boom': '... retailers are reluctant to stock fresh produce that carries a manufacturer's brand. Brunton [of Woolworths] says: 'Do we want to sell, say, Edgell or Nestle fresh produce, or should it be Woolworths fresh produce? Our preference is for the latter, but it is an issue we have to sort out' (N Shoebridge, Fresh food boom demands quick action, *Business Review Weekly,* 18 July 1994, p. 44).

5 According to one review of the operations of the Pick the Tick program, '[a]pproved foods are generally low in fat (or where appropriate have a high polyunsaturated fat: saturated fat ratio) and low in sodium content. They must add some nutritional benefit to the diet' (E Wright, Food endorsement programmes: heartburn for the regulators! *Australian Business Law Review,* 19(5)1991, p. 305).

6 The story of the Australian dietary pyramid is yet to be told, but Professor Marion Nestle's account of the evolution of the United States pyramid shows how science in the public interest is pitted against science in the corporate interest (M Nestle, Dietary advice for the 1990s: the political history of the food guide pyramid, *Caduceus* 9, 1993).

7 On the face of it there does seem some confusion as to whether it was a barbecue or a roast meal.
8 This repositioning as a family meal provider must be paying off given that KFC's busiest day of the year in 1996 was Mother's Day. This signalled that family members thought the product sufficiently okay to treat 'Mum'.
9 Sassen (*The Global City*, Princeton University Press, Princeton, New Jersey 1991) distinguishes producer from consumer services on the basis of spatial distribution: producer services are concentrated in large cities while consumer services are decentralised. Otherwise consumer services remain ill-defined.

8 REASSEMBLING THE CHICKEN: A CULTURAL ECONOMY VIEW OF POWER

1 Luck in the Australian context might constitute a chook stolen from next door.
2 The historic role of traders is indeed fascinating. The capitalist system proper is said to have begun with what is known as proto-industrialisation. In this early form of capital accumulation, merchant capitalists 'put out' raw materials and advanced wages to craftsmen to create products which belonged to the capitalists. This system of production involved the owners of the materials co-ordinating thousands of workers across a district (J Mathews, *Tools of Change: New Technology and the Democratisation of Work,* Pluto Press, Sydney 1989, pp. 11–12). It was the precursor to the system of vertical integration that we see in contemporary agriculture, especially the chicken meat industry and, arguably, is returning due to the power of retail traders to dictate the terms and conditions to producers of chicken meat supply. Today's arrangements see the retail trader 'putting out' contracts to processors who put out contracts to growers. In one sense, the processors are labour overseers for the retailers.
3 Alternative provisioning systems might include food co-operatives, barter networks and 'foodsheds'. This last form of food provisioning is described by J Kloppenburg, J Hendrickson, & G Stevenson, Coming in to the foodshed, *Agriculture and Human Values,* 1996, 13(3): 33–37.
4 The former refers to authority by virtue of the existence of rules and traditions, while the latter refers to claims to rightful authority.
5 There are parallels between my cultural circuits formulation and Weber's notion of economic activity. In *Economy and Society vol.1*, Weber (*The Theory of Social and Economic Organisation*, Free Press, New York, 1947) argues that all activity is at some stage relevant to economic activity, but equally not all activity is determined by the economy. He distinguishes between activities that are 'economic', 'economically relevant' and 'economically determined'. The first category of activity refers to the peaceful means of acquiring control over what he terms utilities. Other forms of activity, such as religious activity are not necessarily bound up with economic activity, but are of relevance for the way in which they influence the needs and propensities by which individuals make use of the utilities. Weber calls this economically relevant activity. Economically determined activities are actions conditioned by economic activity, such as trade unionism. He points out that some activities are both economically determined and relevant, illustrating this with reference to the role of Calvinism in the formation of rational capitalism.

9 THE GLOBAL CHICKEN

1 Preliminary figures for 2001 show that world meat consumption could be broken down in the following manner: pork, 83 158; beef and veal, 48 262; and poultry, 43 241. The figures refer to 1000 metric tons and carcass weight equivalent. The comparative figures since the mid-1990s indicate a relatively rapid narrowing of the gap between beef and poultry (United States Department of

Agriculture Foreign Agriculture Service (March 2002), view date: 28 March 2002, <http://www.fas.usda.gov/dlp/circular/2002/02-03LP>)

2 The latter region's production systems have collapsed with the demise of state socialism (S Tomado, *Safety and Health of Meat, Poultry and Fish Processing Workers,* International Labour Office, Geneva 1997).

3 Buttel argues that 'domestic vertical integration and contracting' in the swine and poultry industries is proof that some agrofood sectors cannot be globalised (F Buttel, Theoretical issues in global agri-food restructuring, in D Burch, R Rickson & G Lawrence (eds) *Globalization and Agri-Food Restructuring: Perspectives from the Australasia Region,* Avebury, Aldershot, 1996, p. 34).

4 The United States, however, does not encourage food self-sufficiency in other countries and provided until recently subsidised exports of poultry to Russia under the heading of aid (Rabobank International, 2001b, *The UK Poultry Industry,* Rabobank International Industry Note, 021–2001, p. 2).

5 When I asked my sister who had worked for KFC in the early 1970s, what she remembered about the experience she said the male customers who would leer and ask her, 'are *you* finger lick'n good?'

6 Whether one can generalise Mintz's argument about whiteness and esteem to chicken meat is doubtful when numerous consumers favour black, grey and yellow chicken. Indeed of five determinants of poultry trade flows the IFC nominated colour differences, and they note the significant trade of 'dark meat poultry parts' to 'countries where consumers have different preferences' (International Finance Corporation, *The World Poultry Industry,* The World Bank, Washington DC 1995, p. 46; Rabobank 2001a, *The Uncertain Path of Trade Liberalisation for the Asian poultry Industry,* Rabobank International Industry Note, 016-2001).

7 P Atkinson (Eating virtue, in A Murcott (ed.) *The Sociology of Food and Eating,* Gower, Aldershot, 1983, pp. 9–17) used this term when describing how food and food mythology can be used to convey what is natural, proper and virtuous.

8 See K Humphery (*Shelf Life: Supermarkets and the Changing Culture of Consumption,* Cambridge University Press, Cambridge 1998, pp. 64–65) for an account of this term.

9 The Rabobank (*The World Poultry Market,* Rabobank, Nederland, 1993) is the only authority I have seen to credit chicken's success with its taste.

REFERENCES

ABARE (1994) *Commodity Statistical Bulletin 1994.* ABARE, Canberra.

Abercrombie, N (1994) Authority and consumer society. In R Keat, N Whiteley & N Abercrombie (eds) *The Authority of the Consumer.* Routledge, London, pp. 43–57.

Alexander, S (1991) *Stephanie's Australia.* Allen & Unwin, Sydney.

—— (1996) *The Cook's Companion.* Viking, Melbourne.

Anonymous (1995) Around the table: Cross-docking. *Supermarket* June, pp. 9–13.

—— (1999) *Eat Better. Live Better.* Coles Media Release, 16 July.

Antonovsky, A (1979) *Health, Stress and Coping.* Jossey-Bass, San Francisco.

Appadurai, A (1986) Introduction: Commodities and the politics of value. In A Appadurai (ed.) *The Social Life of Things.* Cambridge University Press, Cambridge, UK, pp. 3–63.

—— (1990) Disjuncture and difference in the global cultural economy. *Public Culture* 2(2): 1–24.

Arce, A & Marsden, T (1993) The social construction of international food: A new research agenda. *Economic Geography* 69: 293–311.

Atkinson, P (1983) Eating virtue. In A Murcott (ed.) *The Sociology of Food and Eating.* Gower, Aldershot, UK, pp. 9–17.

Aull-Hyde, R, Gempesaw, C & Sundaresan, I (1994) Procurement policies in the US broiler industry: Shall we call them JIT? *Production and Inventory Management Journal* Second Quarter, pp. 11–15.

Australian Bureau of Statistics (1994) *Home Production of Selected Foodstuffs.* Australia, year ended April 1992, ABS, Canberra.

—— (2001) *Yearbook Australia 2001.* ABS, Canberra.

Australian Chicken Meat Federation (n.d.) *Poultry Meat Industry.* ACMF, Sydney.

Australian Competition and Consumer Commission (1999) *ACCC not to Intervene in Bartter/Steggles Poultry Acquisition.* ACCC Media Release.

Australian Consumers Association (1990) The new breed. *Choice* March, pp. 36–37.

—— (1994) Putting fast food to the test. *Choice* April, pp. 7–12.

Australian Food Council (n.d.) Fact Sheet. AFC.

Australian Supermarket Institute and ACNielsen (1998) *Trends in Consumer Behaviour.* ASI/ACNielsen.

Bannerman, C (1998) *Acquired Tastes: Celebrating Australia's Culinary History.* National Library of Australia, Canberra.
Bauman, Z (1988) *Freedom.* Open University Press, Milton Keynes.
—— (1992) *Imitations of Postmodernity.* Routledge, London.
Beardsworth, A & Keil, T (1992) The vegetarian option: Varieties, conversions, motives and careers. *The Sociological Review* 40(2): 253–93.
—— (1997) *Sociology on the Menu.* Routledge, London.
Beasely, C (1994) *Sexual Economyths: Conceiving a Feminist Economics.* Allen & Unwin, Sydney.
Beck, U, Giddens, A & Lash, S (1994) *Reflexive Modernization.* Stanford University Press, Stanford.
Belasco, W (1989/1993) *Appetite for Change: How the Counter Culture Took on the Food Industry.* Cornell University Press, Ithaca.
Bell, N (1990) Introduction. In D Cain *History of the Australian Chicken Meat Industry 1950–1990.* Australian Chicken Meat Federation, Sydney, pp. 9–10.
BIS Shrapnel (1995) *Fast Food in Australia 1995–1997*, 2nd edn. BIS Shrapnel, Sydney.
Bittman, M (1992) *Juggling Time: How Australian Families Use Time.* AGPS, Canberra.
Bittman, M & Mathur, S (1994) Can people buy their way out of housework? Paper presented at the eighth World Congress of Sociology, July, Bieldfeld, Germany.
Bittman, M & Pixley, J (1997) *The Double Life of the Family.* Allen & Unwin, Sydney.
Blackett, Don (1970) "Quo Vadis": The poultry industry in Australia 1970. *Food Technology in Australia* August, pp. 446–47.
Blomley, N (1996) "I'd like to dress her all over": Masculinity, power and retail space. In N Wrigley & M Lowe (eds) *Retailing, Consumption and Capital: Towards the New Retail Geography.* Longman, Harlow, UK, pp. 238–56.
Bonanno A, Busch L, Friedland W, Gouveia L & Mingione E (1994) *From Columbus to ConAgra.* University Press of Kansas, Lawrence, Kansas.
Bottomore, T (1983) (ed.) A *Dictionary of Marxist Thought.* Harvard University Press, Cambridge, Massachusetts. pp. 332–33.
Bourdieu, P (1977) *Outline of a Theory of Practice.* Cambridge University Press, Cambridge, UK.
—— (1984) *Distinction: A Social Critique of the Judgement of Taste.* Routledge, London.
—— (1990) *In Other Words: Essays Towards a Reflexive Sociology.* Polity Press, Cambridge, UK.
Bowler, I (1994) The institutional regulation of uneven development: The case of poultry production in the province of Ontario. *Transactions: Institute of British Geographers* 19: 346–58.
Boyd, W & Watts, M (1997) Agro-industralization just-in-time: The chicken industry and postwar American capitalism. In D Goodman & M Watts (eds) *Globalising Food: Agrarian Questions and Global Restructuring.* Routledge, London, pp. 192–225.
Burch, D & Goss, J (1999) Global sourcing and retail chains: Shifting relationships of production in Australian agri-foods. *Rural Sociology* 64(2): 334–50.
Burch, D, Goss, J & Lawrence, G (1999) *Restructuring Global and Regional Agricultures: Transformations in Australasian Agri-Food Economies and Spaces.* Ashgate, Aldershot, UK.
Burch, D & Pritchard, B (1996) The uneasy transition to globalization: Restructuring of the Australian tomato processing industry. In D Burch, R Rickson & G Lawrence (eds) *Globalisation and Agri-Food Restructuring: Perspectives from the Australasia Region.* Avebury, Aldershot, UK, pp. 107–126.

Burch, D, Rickson, R & Lawrence, G (1996) *Globalisation and Agri-Food Restrucuring: Perspectives from the Australasia Region.* Avebury, Aldershot, UK.

Burgess, J, Campbell, I & McDondald, D (1996) Globalisation and labour standards in Australia. Paper presented at the Conference on Labour Rights and Globalisation: Australia and Asia Pacific, Deakin University, Melbourne.

Buttel, F (1996) Theoretical issues in global agri-food restructuring. In D Burch, R Rickson & G Lawrence (eds) *Globalization and Agri-Food Restructuring: Perspectives from the Australasia Region.* Avebury, Aldershot, UK, pp. 17–44.

Cahn, A (1977) Australians in the early twentieth century. In B Wood (ed.) *Tucker in Australia.* Hill of Content, Melbourne, pp. 53–63.

Cain, D (1990) *History of the Australian Chicken Meat Industry 1950–1990.* Australian Chicken Meat Federation, Sydney.

—— (1996) *South Australian Chicken Meat Industry.* Unpublished Report to the Minister for Primary Industries, March.

Campbell, C (1997) Shopping, pleasure and the sex war. In P Falk & C Campbell (eds) *The Shopping Experience.* Sage, London, pp. 166-176.

Campbell, H & Coombes, B (1999) New Zealand's organic food exports: Current interpretations and new directions in research. In D Burch, J Goss & G Lawrence (eds) *Restructuring Global and Regional Agricultures: Transformations in Australasian Agri-Food Economies and Spaces.* Ashgate, Aldershot, UK, pp. 61–74.

Carson, J (1995) Making fresh food faster. *Australian Farm Journal* September, pp. 44–46.

Casper, C (1996) Homeward bound. *Restaurant Business* 95(10): 165–72.

Chapman, S (1999) Our last line of resistance is the front line. *Weekend Australian* June, pp. 26–27.

Charles, N & Kerr, M (1986) Issues of responsibility and control in the feeding of families. In S Rodmell & A Watt (eds) *The Politics of Health Education: Raising the Issues.* Routledge and Kegan Paul, London, pp. 57–75.

Chicken Meat Research & Development Council (1995) *Annual Report 1994–95.* CMR&D Council, Sydney.

Cockburn, A (1996) A short, meat-oriented history of the world: From Eden to the Mattole. *New Left Review* Jan/Feb 215: 16–42.

Cockburn, C & Ormrod, S (1993) *Gender and Technology in the Making.* Sage, London.

Coles Ltd *Colesanco* issues March 1959; December 1960; December 1961; and June 1962.

Coles Myer Ltd *Annual Reports* 1994, 1995 and 1996.

—— (1995) *A Brief History of the Company.*

Counihan, C (1988) Female identity, food, and power in contemporary Florence. *Anthropological Quarterly* 61(2): 51–62.

Crotty, Patricia (1995) *Good Nutrition? Fact and Fashion in Dietary Advice.* Allen & Unwin, Sydney.

CSIRO (1994) *Information Needs and Concerns in Relation to Food Choice.* CSIRO, Adelaide.

Dairy Farm International Holdings Ltd (1995) *Annual Report 1994.*

Darrall, J (1992) The response of the food chain to healthy eating. *British Food Journal* 94(4): 7–11.

Dawson, John (1995) Food retailing and the food consumer. In DW Marshall (ed.) *Food Choice and The Consumer.* Blackie Academic & Professional, London, pp. 77–104.

DeVault, M (1991) *Feeding the Family: The Social Organization of Caring as Gendered Work.* The University of Chicago Press, Chicago.

Dixon, J (2000) Cooks, chooks and culinary cultures: A cultural economy model for the study of food commodity systems. PhD thesis, Royal Melbourne Institute of Technology.

Dixon, J & Burgess, J (1998) When local elites meet the WTO: Chicken as meat in the sandwich. *Journal of Australian Political Economy* 41: 104–133.

Douglas, M (1997) In defence of shopping. In P Falk & C Campbell (eds) *The Shopping Experience*. Sage, London, pp. 15–30.

Ducatel, K & Blomley, N (1990) Rethinking retail capital. *International Journal of Urban and Regional Research* 14: 207–227.

Duruz, J (1994) Laminex dreams. *Meanjin* 53(1): 99–110.

Eco-Consumer (1995) Kentucky Died Chicken. *Eco-Consumer Newsletter* December, p. 4.

—— (1996) Chicken welfare: Still an issue. *Eco-Consumer Newsletter* February, pp. 1–3.

Elchardus, M (1991), Flexible men and women. The changing temporal organization of work and culture: An empirical analysis. *Social Science Information* 130(4): 701–725.

—— (1994) In praise of rigidity: On temporal and cultural flexibility. *Social Science Information* 33(3): 459–77.

Emerson, R (1990) *The New Economics of Fast Food*. Van Nostrand, New York.

Evers, H & Schrader, H (1994) Introduction. In H Evers & H Schrader (eds) *The Moral Economy of Trade*. Routledge, London, pp. 3–6.

Fairbrother, J (1988) The poultry industry: Technology's child two decades on. *Food Australia* November, pp. 456–62.

—— (1996) Outlook for Poultry. *ABARE Outlook 96 Conference Proceedings*. Australian Bureau of Agriculture and Resource Economics, Canberra.

—— (2001) *Submission to the Department of Foreign Affairs and Trade on Forthcoming Multilateral Trade Negotiations in the World Trade Organization*. The Australian Chicken Meat Federation, June 2001.

Falk, P (1991) Homo culinarius: Towards an historical anthropology of taste. *Social Science Information* 30(4): 757–90.

—— (1994) *Consuming Body*. Sage. London.

Falk, P & Campbell, C (eds) (1997) *The Shopping Experience*. Sage, London.

—— (1997) Introduction. In P Falk & C Campbell (eds) *The Shopping Experience*. Sage, London, pp. 1–14.

Fiddes, N (1991) *Meat: A Natural Symbol*. Routledge, London.

—— (1995) The omnivore's paradox. In D Marshall (ed.) *Food Choice and the Consumer*. Blackie Academic & Professional, London, pp. 131–51.

Fine, B (1994) Towards a political economy of food. *Review of International Political Economy* 1(3): 519–46.

Fine, B, Heasman, M & Wright, J (1996) *Consumption in the Age of Affluence: The World of Food*. Routledge, London.

Fine, B & Leopold, E (1993) *The World of Consumption*. Routledge, London.

Finkelstein, J (1989) *Dining Out: A Sociology of Modern Manners*. Polity, Cambridge, UK.

Fischler, C (1988) Food, self and identity. *Social Science Informatio*n 27: 275–92.

—— (1993) A nutritional cacophony or the crisis of food selection in affluent societies. In P Leathwood, M Horisberger & W James (eds) *For a Better Nutrition in the 21st Century*. Vevey/Raven Press, New York, pp. 57–65.

Fiske, J (1991) *Reading the Popular*. Routledge, London.

Flanagan, B (1995) Fight for the grocery dollar is tacking on a fresh look. *Retail World* December 11–15, p. 17.

Flint, D (1981) Australian eating patterns. In M Wahlquist (ed.) *Food and Nutrition in Australia*. Cassell Australia, Sydney.

Flynn, A & Marsden, T (1992) Food regulation in a period of agricultural retreat: The British experience. *Geoforum* 23: 85–93.
Foord, J, Bowlby, S & Tillsley, C (1996) The changing place of retailer-supplier relations in British retailing. In N Wrigley & M Lowe (eds) *Retailing, Consumption and Capital: Towards the New Retail Geography*. Longman, Harlow, UK, pp. 68–89.
Franklins (n.d.) *The 'No Frills' History*. Dairy Farm International Holdings Ltd.
—— (n.d.) *Background — Franklins Limited*. Dairy Farm International Holdings Ltd.
—— (n.d.) *The one others try to equal*. Dairy Farm International Holdings Ltd.
Friedberg, S (1997) Sustaining trust: Towards a moral economy of food provisioning. Paper presented to the ninety-third annual meeting of American Geographers, April, Texas.
Friedland, W (1984) Commodity systems analysis: An approach to the sociology of agriculture. *Research in Rural Sociology and Development* 1: 221–35.
—— (1994) The new globalization: The case of fresh produce. In A Bonanno, L Busch, W Friedland, L Gouveia & E Mingione (eds) *From Columbus to ConAgra*. University Press of Kansas, Lawrence, Kansas, pp. 210–31.
—— (2001) Reprise on commodity systems methodology. *International Journal of Sociology of Agriculture and Food* 9(1): 82–103.
Friedmann, H (1990) Family wheat farms and third world diets: A paradoxical relationship between unwaged and waged labour. In J Collins & M Gimenz (eds) *Work Without Wages*. State University Press of New York, Albany, pp. 193–213.
—— (1991) Changes in the international division of labor: Agri-food complexes and export agriculture. In W Friedland (ed.) *Towards a New Political Economy of Agriculture*. Westview Press, Boulder, Colarado, pp. 65–93.
—— (1993) After Midas's feast: Alternative food regimes for the future. In P Allen (ed.) *Food for the Future: Conditions and Contradictions of Sustainability*. John Wiley & Sons, New York, pp. 213–33.
—— (1994) Distance and durability: Shaky foundations of the world food economy. In P McMichael (ed.) *The Global Restructuring of Agro-Food Systems*. Cornell University Press, Ithaca, pp. 258–76.
Friedmann, H & McMichael, P (1989) Agriculture and the state system. *Sociologia Ruralis* XXXIX (2): 93–117.
Frow, J (1995) *Cultural Studies and Cultural Value*. Clarendon Press, Oxford.
Fulton, A & Clark, R (1996) Farmer decision making under contract farming in northern Tasmania. In D Burch, R Rickson & G Lawrence (eds) *Globalization and Agri-Food Restructuring: Perspectives from the Australasia Region*. Avebury, Aldershot, UK, pp. 219–38.
Gabriel, Y & Lang, T (1995) *The Unmanageable Consumer*. Sage, London.
Gardner, C & Sheppard, J (1989) *Consuming Passion: The Rise of Retail Culture*. Unwin Hyman, London.
Gawenda, M (1996) What makes Solly run? *The Age Good Weekend* 10 August, pp. 25–31.
Giddens, A (1990) *The Consequences of Modernity*. Stanford University Press, Stanford.
—— (1991) *Modernity and Self-Identity*. Stanford University Press, Stanford.
—— (1992) *The Transformation of Intimacy*. Polity Press, Cambridge, UK.
Gofton, L (1990) Food fears and time famines. In M Ashwell (ed.) Why we eat what we eat. *The British Nutrition Foundation Bulletin* 15(1): 78–95.
Gofton, L & Ness, M (1991) Twin trends: Health and convenience in food change or who killed the lazy housewife. *British Food Journal* 93(7): 17–23.
Goodman, D & Redclift, M (1991) *Refashioning Nature: Food, Ecology and Culture*. Routledge, London.

Goodman, D & Watts, M (1997) *Globalising Food: Agrarian Questions and Global Restructuring*. Routledge, London.

Grant, R (1993) Against the grain: Agricultural trade policies of the US, the European Community and Japan at the GATT. *Political Geography* 12(3): 247–62.

Gray, I, Lawrence, G & Dunn, T (1993) Coping with change: Australian farmers in the 1990s. Centre for Rural Social Research, Charles Sturt University, Wagga Wagga, NSW.

Gregory, C & Altman, J (1989) *Observing the Economy*. Routledge, London.

Halperin, R (1994) *Cultural Economies: Past and Present*. University of Texas Press, Austin.

Harris, M (1986) *Good to Eat: Riddles of Food and Culture*. Allen & Unwin, London.

Harvey, D (1989) *The Condition of Postmodernity*. Basil Blackwell, Oxford.

—— (1996) *Justice, Nature and the Geography of Distance*. Blackwell Publishers, Cambridge, Massachusetts.

Harvey, M (1998) UK supermarkets: New product and labour market segmentation and the restructuring of the supply-demand matrix. Paper presented to the International Working Party on Labour Market Segmentation Conference, Trento, Italy.

Hazen, J (1994) *Chicken Soup Book: Old and New Recipes From Around the World*. Chronicle, USA.

Held, D & Thompson, J (1989) *Social Theory of Modern Societies: Anthony Giddens and his Critics*. Cambridge University Press, Cambridge, UK.

Hindess, B (1996) *Discourses of Power: From Hobbes to Foucault*. Blackwell Publishers, Oxford.

Hughes, Alexandra (1996) Forging new cultures of food retailer-manufacturer relations? In N Wrigley & M Lowe (eds) *Retailing, Consumption and Capital: Towards the New Retail Geography*. Longman, Harlow, UK, pp. 90–115.

Humphery, K (1998) *Shelf Life: Supermarkets and the Changing Culture of Consumption*. Cambridge University Press, Cambridge, UK.

—— (1995) Talking shop. *Arena Magazine* October/November, pp. 30–34.

Inkson, K & Cammock, P (1988) The meat-freezing industry in New Zealand. In E Willis (ed.) *Technology and the Labour Process*. Allen & Unwin, Sydney, pp. 68–80.

Instate Pty Ltd (1997) *Asian Export Opportunities for Chicken Meat*. Australian Chicken Meat Federation, Sydney.

International Finance Corporation (IFC) (1995) *The World Poultry Industry*. The World Bank, Washington DC.

Ironmonger, D (1989) Australian households: A $90 billion industry. Research Discussion Paper, Centre for Applied Research on the Future, University of Melbourne.

Joint Select Committee on the Retailing Sector (1999) *Fair Market or Market Failure: A Review of Australia's Retailing* Sector. Parliament of Australia, Canberra.

Jussaume, R (1991) Culture and food consumption: Is the world converging on McDonald's? *Sociological Practice Review* 2(1): 49–58.

Keat, R (1994) Scepticism, authority and the market. In R Keat, N Whiteley & N Abercrombie (eds) *The Authority of the Consumer*. Routledge, London, pp. 23–42.

Kim, C & Curry, J (1993), Fordism. Flexible specialization and agri-industrial restructuring: The case of the US broiler industry. *Sociologia Ruralis* XXXiiI(1): 61–80.

Kingston, B (1994) *Basket, Bag and Trolley: A History of Shopping in Australia*. Oxford University Press, Melbourne.

Kloppenburg, J, Hendrickson, J & Stevenson, G (1996) Coming in to the foodshed. *Agriculture and Human Values* 13(3): 33–37.
Koptytoff, I (1986) The cultural biography of things: Commoditization as process. In A Appadurai (ed.) *The Social Life of Things.* Cambridge University Press, Cambridge, UK, pp. 64–91.
Larkin, J & Heilbron, S (1997) *The Australian Chicken Meat Industry: International Benchmarking Study.* Instate Pty Ltd, Sydney, and SG Heilbron, Melbourne.
Larkin, T (1991) *The Australian Poultry Industry: Economic Structure and the Impact of World Poultry Trade Developments.* JT Larkin and Associates, Canberra.
Lash, S (1990) *Sociology of Postmodernism.* Routledge, London.
Lash, S & Urry, J (1994) *Economies of Signs and Space.* Sage, London.
Lawrence, G (1987) *Capitalism and the Countryside: The Rural Crisis in Australia.* Pluto Press, Sydney.
Lawrence, G & Vanclay, F (1994) Agricultural change in the semiperiphery: The Murray-Darling Basin, Australia. In P McMichael (ed.) *The Global Restructuring of Agro-Food Systems.* Cornell University Press, Ithaca, pp. 76–103.
Lee, M (1994a) Worth crowing about! *Supermarket* September, pp. 29–30.
(1994b) Getting a slice of the action. *Supermarket* August, pp. 18–19.
Le Heron, R & Roche, M (1996) Eco-commodity systems: Historical geographies of context, articulation and embeddedness under capitalism. In D Burch, R Rickson & G Lawrence (eds) *Globalization and Agri-Food Restructuring: Perspectives from the Australasia region.* Avebury, Aldershot, UK, pp. 73–90.
Leopold, M (1985) The transnational food companies and their global strategies. *International Journal of Social Science* 37(3): 315–30.
Lester, I (1994) *Australia's Food and Nutrition.* AGPS, Canberra.
Levenstein, H (1988) *Revolution at the Table: The Transformation of the American Diet.* Oxford University Press, Oxford.
—— (1993) *Paradox of Plenty.* Oxford University Press, Oxford.
Levi-Strauss, C (1978) *The Origin of Table Manners: Introduction to a Science of Mythology*, Vol. 3. Jonathan Cape, London.
Lipietz, A (1987) *Mirages and Miracles: The Crisis of Global Fordism.* Pluto, London.
Lowe, Michelle & Wrigley, Neil (1996) Towards the new retail geography. In N Wrigley & M Lowe (eds) *Retailing, Consumption and Capital: Towards the New Retail Geography.* Longman, Harlow, UK, pp. 3–30.
Lukes, S (1974) *Power: A Radical View.* Macmillan, London.
Lupton, D (1996) *Food, the Body and the Self.* Sage, London.
Luxton, M (1980) *More than a Labour of Love.* The Women's Press, Toronto.
Lyons, K (1996) Agro-industrialization and social change within the Australian context: A case study of the fast food industry. In D Burch, R Rickson & G Lawrence (eds) *Globalization and Agri-Food Restructuring: Perspectives from the Australasia Region.* Avebury, Aldershot, UK, pp. 239–50.
McIntosh, W & Zey, M (1989) Women as gatekeepers of food consumption: A sociological critique. *Food and Foodways* 3(4): 317–32.
McLellan, D (trans.)(1973) *Marx's 'Grundrisse'.* Paladin, St Albans, UK.
McMichael, P (1993) Agro-food restructuring in the Pacific Rim: A comparative-international perspective on Japan, S. Korea, the US, Australia and Thailand. In R Palat (ed.) *Pacific Asia and the Future of the World System.* Greenwood Press, Westport, Connecticut.
McMichael, P (1994) (ed.) *The Global Restructuring of Agro-Food Systems.* Cornell University Press, Ithaca.
—— (1996) Globalization: Myths and realities. *Rural Sociology* 61(1): 25–55.

Mackay, H (1992) *The Mackay Report. Food.* Self-published, Sydney.

Marsden, T & Little, J (1990) Introduction. In T Marsden & J Little *Political, Social and Economic Perspectives on the International Food System.* Avebury, Aldershot, UK, pp. 1–15.

Marsden, T & Wrigley, N (1996) Retailing, the food system and the regulatory state. In N Wrigley & M Lowe (eds) *Retailing, Consumption and Capital: Towards the New Retail Geography.* Longman, Harlow, UK, pp. 33–47.

Mathews, J (1989) *Tools of Change: New Technology and the Democratisation of Work.* Pluto Press, Sydney.

—— (1994) *Catching the Wave. Workplace Reform in Australia.* Allen & Unwin, Sydney.

Mencken, W (1996) Women get their chance to run the store. *Supermarket* February, p. 11.

Mennell, S (1985 & 1996) *All Manners of Food: Eating and Taste in England and France from the Middle Ages to the Present.* Basil Blackwell, Oxford.

Mennell, S, Murcott, A & van Otterloo, A (1992) *The Sociology of Food: Eating, Diet and Culture.* Sage, London.

Miller, D (1995) Consumption as the vanguard of history. In D Miller (ed.) *Acknowledging Consumption: A Review of New Studies.* Routledge, London, pp. 1–57.

Miller, P & Rose, N (1997) Mobilizing the consumer: Assembling the subject of consumption. *Theory, Culture & Society* 14(1): 1–36.

Miller, T (1993) *The Well-Tempered Self.* John Hopkins University Press, Baltimore, Maryland.

Mintz, S (1985) *Sweetness and Power: The Place of Sugar in Modern History.* Penguin Books, New York.

—— (1994) Eating and being: What food means. In B Harriss-White (ed.) *Food: Multidisciplinary Perspectives.* Basil Blackwell, Cambridge, UK, pp. 102–115.

—— (1996) *Tasting Food, Tasting Freedom.* Beacon Books, Boston.

Morris, M (1988) Things to do with shopping centres. In S Sheridan (ed.) *Grafts: Feminist Cultural Criticism.* Verso, London, pp. 193–225.

Murcott, A (1982) On the social significance of the 'cooked dinner' in South Wales. *Social Science Information* 21(4/5): 677–96.

—— (1986) "It's a pleasure to cook for him": Food, mealtimes and gender in some South Wales households. In E Gamarnikow (ed.) *The Public and the Private.* Gower, Aldershot, UK, pp. 62–77.

Murdoch, J (1994) Some comments on 'nature' and 'society' in the political economy of food. *Review of International Political Economy* 1: 571–77.

Murray, R (1989) Fordism and post-Fordism. In S Hall & M Jacques (eds) *New Times: The Changing Face of Politics in the 1990s.* Lawrence and Wishart, London.

Nestle, M (1993) Dietary advice for the 1990s: The political history of the food guide pyramid. *Caduceus* 9: 136–53.

—— (1995) Dietary guidance for the 21st century: New approaches. *Journal of Nutrition Education* 27: 272–75.

NSW Farmers (1998) *NSW Farmers Challenge Supermarkets' Power.* NSW Farmers Press Release, 13 October.

Occupational Health and Safety Authority (n.d.) *Manual Handling and Noise in the Poultry Processing Industry.* Department of Business and Employment, Melbourne.

Parsons, H (1996) Supermarkets and the supply of fresh fruit and vegetables in Australia: Implications for wholesale markets. In D Burch, R Rickson & G Lawrence (eds) *Globalization and Agri-Food Restructuring: Perspectives from the Australasia Region.* Avebury, Aldershot, UK, pp. 251–70.

Pelto, G & Pelto, P (1983) Diet and delocalization: Dietary changes since 1750. In R Rotberg & T Rabb (eds) *Hunger and History: The Impact of Changing Food Production and Consumption Patterns on Society*. Cambridge University Press, Cambridge, UK, pp. 309–330.

PepsiCo Restaurants International (n.d.) How KFC increased its market share through an integrated marketing approach including celebrity endorsement for the launch of 'TenderRoast'.

—— (n.d.) Launch and ongoing promotion of "TenderRoast".

—— (n.d.) Launch and ongoing promotion of "Kentucky BBQ".

—— (n.d.) Fact Sheets PepsiCo Operations; KFC Mission Statement; Guildford — where it all began for KFC in Australia; KFC world-wide operations; KFC History Colonel Sanders; KFC Australia; General briefing employment & investment program by KFC.

—— (1994) A typical investor looks us over. *PepsiCo Annual Report*.

—— (1995) Determined to grow. *PepsiCo Annual Report*.

Piore, M & Sabel, C (1984) *The Second Industrial Divide*. Basic Books, New York.

Polanyi, K (1944) *The Great Transformation*. Holt, Rinehart and Winston, New York.

Poultry Grower News (1994–95) Various issues of the US Poultry Growers newsletters, 1994–1995.

Prices Surveillance Authority (1986) *Inquiry in Relation to the Table Chicken Industry*. Report No 7, Prices Surveillance Authority.

Pritchard, B (1995) Foreign ownership of Australian food processing: The 1995 sale of Pacific Dunlop food division. *Journal of Australian Political Economy* 36: 26–47.

—— (1996) Restructuring in the Australian dairy industry: The reconstruction and reinvention of co-operatives. In D Burch, R Rickson & G Lawrence (eds) *Globalization and Agri-Food Restructuring: Perspectives from the Australasia Region*. Avebury, Aldershot, UK, pp. 139–52.

Probert, B (1994) Globalisation, economic restructuring and the state. In S Bell & B Head *State, Economy and Public Policy in Australia*. Oxford University Press, Oxford, pp. 98–118.

Probyn, E (1998) *Mc*-Identities: Food and the familial citizen. *Theory, Culture & Society* 15(2): 155–173.

Rabobank International (1993) *The World Poultry Market*. Rabobank, The Netherlands.

Rabobank International (2001a) *The Uncertain Path of Trade Liberalisation for the Asian Poultry Industry*. Rabobank International Industry Note 016-2001.

Rabobank International (2001b) *The UK Poultry Industry*. Rabobank International Industry Note 021-2001.

Reeders, E (1988) The fast food industry. In E Willis (ed.) *Technology and the Labour Process*. Allen & Unwin, Sydney, pp. 142–54.

Reekie, G (1993) *Temptations: Sex, Religion and the Department Store*. Allen & Unwin, Sydney

Ripe, C (1993) *Goodbye Culinary Cringe*. Allen & Unwin, Sydney.

Robins, K (1994) Forces of consumption: From the symbolic to the psychotic. *Media, Culture & Society* 16: 449–68.

Root, W (1980) *Food: An Authoritative and Visual History and Dictionary of the Foods of the World*. Simon and Schuster, New York.

Ryan, S & Burgess, J (1996) The Supermarket Co. In J Burgess, P Keogh, D Macdonald, G Morgan, G Strachan & S Ryan (eds) *Enterprise Bargaining in Three Female Dominated Workplaces in the Hunter: Processes, Participation and Outcomes*. Employment Studies Centre Working Paper Series No. 26, University of Newcastle, NSW.

Safeway (1995) Understanding Safeway. Module Three — Interactive Workshop, 13 October 1995.

Sanderson, S (1986) The emergence of the 'world steer': Internationalization and foreign domination in Latin American cattle production. In F Tullis & W Hollist (eds) *Food, the State, and International Political Economy.* University of Nebraska Press, Lincoln, pp. 123–48.

Santich, B (1995a) 'It's a chore!' Women's attitudes towards cooking. *Australian Journal of Nutrition and Dietetics* 52(1): 11–13.

—— (1995b) *What the Doctors Ordered: 150 Years of Dietary Advice in Australia.* Hyland House, Melbourne.

Sassen, S (1991) *The Global City.* Princeton University Press, Princeton, New Jersey.

Scott, V & Worsely, A (1994) Ticks, claims, tables and food groups: A comparison for nutrition labelling. *Health Promotion International* 9 (1): 27–37.

Seccombe, W (1986) Patriarchy stabilized: The construction of the male breadwinner wage norm in nineteenth century Britain. *Social History* 11: 53–76.

Senate Select Committee on Animal Welfare (1990) *Intensive Livestock Production.* AGPS, Canberra.

Senker, J (1988) *A Taste for Innovation: British Supermarkets' Influence on Food Manufacturers.* Horton Publishing, Bradford, UK.

Sennett, R (1980) *Authority.* Alfred A Knopf, New York

Shackleton, R (1996) Retailer internationalization: A culturally constructed phenomenon. In N Wrigley & M Lowe (eds) *Retailing, Consumption and Capital: Towards the New Retail Geography.* Longman, Harlow, UK, pp. 137–56.

Shoebridge, N (1993) *The Secrets of Successful Marketing.* The Text Publishing Company, Melbourne.

—— (1994) Fresh food boom demands quick action. *Business Review Weekly* 18 July, pp. 40–44.

—— (1995) Well-done beef campaign manages a rare feat. *Business Review Weekly* 4 September, pp. 48–52.

—— (1996) KFC finds real meals much more satisfying. *Business Review Weekly* 19 February, pp. 64–67.

—— (1998) Hey, big spenders — Australia's top advertisers dig much deeper. *Business Review Weekly* 2 March, pp. 68–71.

Sindall, C, Wright, J & O'Dea, K (1994) Food production and human nutrition: The impact of health messages. A public health perspective. In *Proceedings of the Nutrition Society of Australia* 18.

Silverstone, R, Hirsch, E & Morley, D (1992) Information and communication technologies and the moral economy of the household. In R Silverstone & E Hirsch (eds) *Consuming Technologies: Media and Information in Domestic Spaces.* Routledge, London, pp. 15–31.

Skurray, G & Newell, G (1993) Food consumption in Australia 1970–1990. *Food Australia* 45(9): 434–38.

Sokolov, R (1991) *Why We Eat What We Eat: How the Encounter Between the New World and the Old Changed the Way Everyone on the Planet Eats.* Summit Books, New York.

Stanton, R (1987) Healthy eating featuring chicken. *New Idea* November Supplement.

Steggles (1996) The fresh chicken market: Trends & dynamics. Steggles Internal Document.

Strickland, K (1996) Fast food chains latch on to health. *The Australian* 5 February, p. 3.

Symons, M (1982) *One Continuous Picnic: A History of Eating in Australia.* Duck Press, Adelaide.

Tansey, G & Worsley, T (1995) *The Food System: A Guide.* Earthscan Publications, London.

Taylor, M (1994–95) Observations on the broiler industries in the USA and the UK with particular reference to the relations between growers and processors. *Appendix to the Seventh Annual Report on the Operations and Activities of the Victorian Broiler Industry Negotiation Committee*, VBINC, Melbourne.

Thompson, E P (1968) *The Making of the English Working Class.* Penguin Books, Harmondsworth, UK.

—— (1993) *Customs in Common.* Penguin Books, London.

Tomoda, S (1997) *Safety and Health of Meat, Poultry and Fish Processing Workers.* International Labour Office, Geneva.

Turner, B (1984) *The Body and Society: Explorations in Social Theory.* Basil Blackwell, Oxford.

Turner, C (1977) The Australian National Food Pattern. In B Wood (ed.) *Tucker in Australia.* Hill of Content, Melbourne, pp. 64–75.

Ufkes, F (1993) Trade liberalization, agro-food politics and the globalization of agriculture. *Political Geography* 12(3): 215–31.

United States Department of Agriculture Foreign Agriculture Service (2002) Livestock and Poultry World Markets and Trade, USDA Foreign Agriculture Service, viewed 28 March 2002, <http://www.fas.usda.gov/dlp/circular/2002/02-03LP/>

Van Gramberg, B (1999) Women, industrial relations and public policy. In L Hancock (ed.) *Women, Public Policy and the State.* Macmillan Education Australia, Melbourne, pp. 99–113.

Visser, M (1986) *Much Depends upon Dinner.* Penguin Books, London.

Warde, A (1992) Notes on the relationship between production and consumption. In R Burrows & C Marsh (eds) *Consumption and Class: Divisions and Change.* Macmillan, London, pp. 15–31.

—— (1994) Consumers, identity and belonging: Reflections on some theses of Zygmunt Bauman. In R Keat, N Whiteley & N Abercrombie (eds) *The Authority of the Consumer.* Routledge, London, pp. 58–73.

—— (1997) *Consumption, Food and Taste.* Sage, London.

Waring, M (1988) *Counting for Nothing: What Men Value and What Women are Worth.* Allen & Unwin, Wellington, New Zealand.

Waters, M (1995) *Globalization.* Routledge, London.

Watts, M & Goodman, D (1997) Agrarian questions. In D Goodman & M Watts *Globalising Food.* Routledge, London, pp. 1–32.

Weber, M (1947) *The Theory of Social and Economic Organisation.* Free Press, New York.

Whatmore, S, Munton, R, Little, J & Marsden, T (1987) Towards a typology of farm businesses in contemporary British agriculture. *Sociologia Ruralis* 27(10): 21–37.

Whit, W (1995) *Food and Society: A Sociological Approach.* General Hall, New York.

Whitehead, A (1994) Food symbolism, gender power and the family. In B Harriss-White (ed.) *Food: Multidisciplinary Perspectives.* Basil Blackwell, Cambridge, UK, pp. 116–29.

Williams, M (1997) The political economy of meat: food, culture and identity. Paper presented to the Annual Convention of the International Studies Association, March, Toronto.

Wood, B (1977) *Tucker in Australia.* Hill of Content, Melbourne.

—— (1977) Shopping and eating in Melbourne. In B Wood *Tucker in Australia.* Hill of Content, Melbourne, pp. 76–78.

Woolworths Limited *Annual Reports* 1995 and 1996.

Woolworths Ltd (n.d.) *70th Anniversary 1924–1994. Woolies News* Special Souvenir Edition.

—— (n.d.) *A Brief History of Safeway.*

Wray, D (1995) Poultry consumption: No end in sight. *Misset World Poultry* 11(5): 10–15.

Wright, E (1991) Food endorsement programmes —heartburn for the regulators! *Australian Business Law Review* 19(5): 303–324.

Wrigley, N & Lowe, M (1996) *Retailing, Consumption and Capital: Towards the New Retail Geography*. Longman, Harlow, UK.

Zola, E (1992) *The Ladies' Paradise*. University of California Press, Berkeley.

Zukin, S (1991) *Landscapes of Power: From Detroit to Disney World*. University of California Press, Berkeley.

INDEX